on the edge

char miller

on the edge

water, immigration, and politics in the southwest

Trinity University Press | SAN ANTONIO

Cover design by Rebecca Lown
Book design by BookMatters, Berkeley
Cover illustration: Rain on parched, cracked ground,
©iStockphoto.com/cjp

Trinity University Press strives to produce its books using methods and materials in an environmentally sensitive manner. We favor working with manufacturers that practice sustainable management of all natural resources, produce paper using recycled stock, and manage forests with the best possible practices for people, biodiversity, and sustainability. The press is a member of the Green Press Initiative, a nonprofit program dedicated to supporting publishers in their efforts to reduce their impacts on endangered forests, climate change, and forest-dependent communities.

The paper used in this publication meets the minimum requirements of the American National Standard for Information Sciences—Permanence of Paper for Printed Library Materials, ANSI 39.48-1992.

ISBN 978-1-59534-147-1 (paper)
ISBN 978-1-59534-148-8 (ebook)

CIP data on file at the Library of Congress.

17 16 15 14 13 | 5 4 3 2 1

For Hutze and Sam

Contents

Introduction
Moving in Place

life and the memory of it so compressed
they've turned into each other. Which is which?
— *Elizabeth Bishop, "Poem"*

We live in place. That claim is as academic as it is grounded, for we inhabit some places that are imagined, constructed, or recalled and others that are physical, precise, and mutable. Some are landscapes we have consciously planted ourselves in, and others contain dimensions we are only dimly aware of. Some are contested and controversial, some privileged or poor, others contented. We have invented words to describe the ideal setting for a more benign life (Eden) and its opposite (Hell, for one). Because we always tell stories about these places, however beneficent, and about our travels between them; because these tales are often set within the biographical arc of our lives; and because they serve as touchstones to demarcate, highlight, or illustrate our experiences, we know these narratives to be personal and have the potential to move us. To live in place is to live in many places.

To comprehend the intertwined and layered worlds

we move through, we have invented a form of storytelling we call history. Its tools are diverse, from the autobiographical to the historiographical. Its ambitions can be macro or micro in scope, and its sources draw on material culture, oral history, literary leavings, documentary evidence, and memory. But always these objects, words, images, and remembrances matter because they are contingent; they are relevant and revealing because of their relationship to other times and places (and the spaces in between).

"Space and place are a basic component of the lived world," geographer Yi-Fu Tuan has observed, but we often take these elements for granted and so do not always comprehend how we live within them. Only when "we think about them . . . [do] they assume unexpected meanings and raise questions we had not thought to ask."[1] That kind of introspection frames writer Barry Lopez's appeal: to "hear the unembodied call of a place, that numinous voice, one has to wait for it to speak through the harmony of its features—the soughing of the wind across it, its upward reach against a clear night sky, its fragrance after a rain. One must wait for the moment when the thing—the hill, the tarn, the lunette, the kiss tank, the caliche flat, the bajada—ceases to be a thing and becomes something that knows we are there."[2]

Yet our knowing we are there is also part of the process of reclaiming a sense of place. Take the garden. Not the biblical one but the one Michael Pollan began to dig in rural northeastern Connecticut, a verdant, rocky landscape not far from where I grew up. As he put shovel to sod, cutting through the lawn that surrounded his derelict farm and "had spread over this place like a lid on its agricultural past," he was amazed by what he unearthed—"the skeletons of dogs and deer, a plow, several decomposing tractor tires, a couple of rusty bowie knives, children's toys." This detritus was archaeological in significance. "A place like this is a kind of palimpsest," he wrote, "and much of our gardening has been a process of laying bare the marks of earlier hands on this land."[3]

How those earlier presences speak to us, how we recover their import and incorporate them into our consciousness, is another matter. So a baffled Matthew Halland mused. Standing on a busy London thoroughfare, the protagonist of Penelope Lively's novel *City of the Mind* realizes he has never noticed the meanings embedded in names of his hometown's many districts, streets, and bus routes: "[London] mutters still in Anglo-Saxon; it remembers the hills that have become the Neasden and Islington and Hendon, the marshy islands of Battersea and Bermondsey. The ghost of another topography lingers; the uplands and the streams, the woodland and fords are inscribed still on the London Streetfinder, on the ubiquitous geometry of the Underground map, in the destinations of buses. The Fleet River, its last physical trickle locked away underground in a cast iron pipe, leaves its name defiant and untamed upon the surface. The whole place is one babble of allusions, all chronology subsumed into the distortions and mutations of today, so that in the end what is visible and what is uttered are complementary."[4]

These apparitional presences may mold our recognition of place. Indeed our attentiveness to them is how we *create* a sense of place, geographer Tuan suggests: "What begins as undifferentiated space becomes place as we get to know it better and endow it with value." That is what I have come to recognize about a windswept sandy spit of land that has been a touchstone for much of my life. Chappaquiddick Island, seven miles off the coast of Cape Cod in Massachusetts, is perhaps most famous as the site of the late Ted Kennedy's automobile accident that killed a young staffer and sank his political ambitions. But to those who know it as Chappy—a diminutive that banishes the ancient indigenous claim to the land and also distinguishes between contemporaries who are in or out of the know—the isle's primal significance is its ongoing, shape-shifting instability.

By the good graces of the Atlantic Ocean, Chappaquiddick is often not an island at all; rather it is linked to the larger Martha's Vineyard Island by a long stretch of tidal-sculpted shoreline. Not that I knew this as a child, when during our summer-long residences my sisters and I swam in Chappaquiddick's cool waters; sailed, fished, and crabbed on its waterways, wetlands, and estuaries; and baked ourselves on its fine-grained beaches. It turns out that the very sand was a clue to the island's protean character. In late adolescence I pulled a well-thumbed edition of Barbara Blau Chamberlain's *These Fragile Outposts* (1964) from my mother's bookshelf. In it, Chamberlain describes how the Laurentide ice sheet scoured the surfaces of Cape Cod, Martha's Vineyard, and Nantucket, leaving behind a formidable terminal moraine when it finally retreated more than 18,000 years ago. On the well-worn remnants of this gravelly expanse my siblings and I not infrequently stubbed our toes. Ever since, to walk these beaches has been to be in commune with this ancient, elemental presence.

The varied ways we give meaning to the landscapes we occupy—historical, personal, physical—are also the organizing principles of *On the Edge*, which contains essays about the American Southwest, a region I have known, loved, and misunderstood. The chapters navigate between its two coasts, the Gulf of Mexico and the Pacific; their often turbulent waters in turn help frame some of the book's geographic orientation and narrative structure. Watching hurricanes slam into New Orleans or tear through Texas beach communities, witnessing winter storms and churning tides undercut concrete barriers protecting West Coast harbor and beach, and remembering the historic runs of steelhead trout as they surged out of the ocean and up Southern California's rivers and creeks to spawn are all ways into this story about the difficulties species face in trying to find (and hold) their niche in this wider region's complicated nature.

There are other points of access, including three locales that have been central to my professional life as a teacher, writer, and historian: San Antonio and Los Angeles, and the contested geopolitical frontier they bookmark and are often identified by, the southwestern borderlands. I lived in San Antonio between 1981 and 2007, and I attended college in Southern California in the 1970s and returned to teach in the same community in 2007. Because life in these two burgeoning, minority-majority cities is routinely, deeply influenced by the transborder migrations of people, culture, foodways, and ideas, you cannot reside in either without turning your eyes south and taking into account the transformative power of the U.S.-Mexico boundary. The struggles of the border are those of its urban hubs.

Learning to see these tangled connections requires knowing something about the locations themselves, about their histories, their sites and situations and the often hidden interplay between them. The first thing to know—and I did not know it until I lived there—is how powerful the Spanish legacy has been in shaping the region's look, feel, shape, and sound. The first written words about this terrain were often in that imported European language, a linguistic paradigm and imperial perspective that managed to erase those that came before. Since the late nineteenth century the remnants of New Spain's eighteenth-century built landscape—its missions, presidios, and villas, and the Alamo!—have emerged as the building blocks of the modern tourist industry (colonialism is just as big a sell here as Williamsburg or Plimoth Plantation is in the East). By the 1920s, white adobe, decorative terra-cotta, arched doorways, and open courtyards filled with bougainvillea had become staples in the ubiquitous mission style that has flourished ever since in San Antonio, El Paso, Tucson, Los Angeles, Laredo, and San Diego, giving texture to residences, apartment buildings, hotels, and civic centers. These architectural motifs and aesthetic sensibilities come with a sharp edge: the Anglo elite appropriated them, curiously enough, to assert their claim over the landscape Spain once dominated and to marginalize contemporary Mexican American communities and their aspirations. There is bad blood in those red-tile roofs.[5]

I initially missed other manifestations of how deeply the past can infiltrate the present, influencing its physical dimensions and inflecting its movement. Spend time in Main Plaza in the heart of downtown San Antonio or stroll along Los Angeles's Olvera Street. It doesn't take long before you begin to comprehend how eighteenth-century Spanish urban planners designed the central core to be communal and pedestrian; these elements are still visible amid modern auto-centric streetscapes. Freeways have not bulldozed all evidence of the nineteenth-century transportation grid either. In the early 1990s, while driving along San Antonio's Loop 1604, I realized I did not really understand this late-twentieth-century freeway. Its

100-mile circumference gives form to the sprawling suburbanization that characterizes the massive South Texas city. The associated malls, subdivisions, and fast-food infrastructure, which fan out from every entrance and exit, seemed to constitute a new urban landscape. Then, after ten years of living in the Alamo City, I finally noticed the many streets that intersected with the high-speed highway—Somerset, Pearsall, and Pleasanton Roads; Old Corpus Christi and Old Nacogdoches; Bulverde, Bandera, and Fredericksburg. All are named for towns that defined San Antonio's nineteenth-century hinterland. Some are nestled in the folds of the Edwards Plateau to the north and west, and others dot the coastal plains that fall away to the south and east. All were farm-to-market conduits to herd cattle, goats, and turkeys into local stockyards; down them too rolled horse-drawn wagons and mule carts hauling bales of cotton, cords of firewood, and mounds of produce for sale in the city's open-air markets on Main and Military Plazas. This older network, its economic energy and environmental import, suddenly came alive, forcing me to reconsider how I taught and wrote about the dynamic reciprocity between then and now.

Roadways also figure crucially in my (re)integration into another home ground—Claremont, California, located forty miles east of Los Angeles. The community of 35,000 sits atop an alluvial fan that flows out of the Mount San Antonio watershed, soil that was perfect for the citrus production powering the local economy in the early twentieth century. The town is now an academic arcadia, housing two graduate institutions and a consortium of five undergraduate colleges. In the early 1970s my wife, Judi, and I attended the youngest of them, Pitzer College (est. 1963); thirty years later our son matriculated at the oldest, Pomona College (est. 1889); and in 2007 Judi and I returned so that I could teach at Pomona as a visiting professor, a temporary position that in 2009 became full-time. Our first semester back was disorienting; having lived in this leafy suburb during three different occasions—as students, parents, and teacher—who were we? Rounding a particular corner in the village, hiking up Icehouse Canyon, or entering a classroom where we had once studied let loose a stream of memories whose location in time was unclear, confounding. We were in place but also out of it.

What remained most palpable, most immediately familiar, was Claremont's self-conscious pattern of naming streets. That task had fallen to its late-nineteenth-century founders, who also established Pomona College, using the school to anchor their land development schemes. Many hailed from New England and hungered to replicate the region's intellectual affectations. So they and succeeding generations named the new north-south streets after prestigious universities and colleges they had attended, or wished they had. Running parallel with the central axis defined by College Avenue are Harvard and Yale, and Dartmouth and Columbia. As the city expanded this conceit rippled outward: Berkeley and Stanford (these arch rivals even intersect); Oxford and Cambridge; Cornell, Grinnell, and Oberlin. A section devoted to women's colleges, including Radcliffe, Scripps, Vassar, and Simmons, are of a piece with a bastion of military schools (West Point, Annapolis, and Citadel) and a clutch of southern temples of higher learning (Vanderbilt, Emory, and Duke). For all its symbolic value, Claremont's insistence on scoring its educative mission into its urban form carries a psychological undertone. Could this far-West outpost ever measure up to what it considered to be the East's greater cultural cachet? Somehow this communal anxiety helped us negotiate our homecoming.

No place is less stable and homelike, though, than the U.S.-Mexico border, a reflection of the U.S. government's attempt since 2005 to criminalize the landscape by militarizing it as a place. The long shadow the infamous border wall casts, like the pernicious use of high-tech surveillance equipment, high-speed vehicles, and well-armed Border Patrol agents, is designed to terrorize the terrain, or rather to unsettle our perceptions of it. You need only encounter the three-layer fencing, the high-intensity spotlights and whiz-bang sensors, to recognize that this infrastructure is designed to keep out undocumented migrants and announce that the space itself is illegal. We are simultaneously attacking the people and the place, tearing asunder the land—urban and rural—that has sheltered humans as well as ocelot and jaguarundi. This is a troubling sign of a society that seems more brittle, less nimble; more fearful, less welcoming. Compare this overt hostility toward migrants, regardless of their status, to the more generous and benign message evoked in Emma Lazarus's 1883 poem "The New Colossus," which is emblazoned on the base of the Statute of Liberty: "Give me your tired, your poor, your huddled masses yearning to breathe free." We share none of that earlier generation's confidence or conviction about the nation's pride of place.

How Americans make sense of themselves in the larger world is one more reason to write about my experiences living in San Antonio, Los Angeles, and the borderlands that stretch between them. These southwestern locales are at once the stimulus and subject of my essays, trigger mechanisms that remind me of episodes whose meanings I might use to thicken the texture of these stories, real and remembered.

This interplay of autobiography and environment has political implications, too. I could not have known that moving to the Southwest in 1981 would profoundly impact my sense of citizenship, or that it would lead me to begin speaking to and writing for the larger public about the environmental pressures, judicial struggles, social injustices, and economic disparities that have troubled the communities I have resided in.

Consider the disappearance of houses and landscapes. Bulldozers tearing into the built fabric of a Texas neighborhood, like the earthmoving equipment that was mobilized to construct an armored border wall that has sealed off the Southwest—Homeland Security calls it "tactical infrastructure"—represents the loss of historic relationships, an attenuation of identity and integrity that necessarily leads to a diminution of place. This decline is even more affective with the death of parents and relatives whose powerful presence in our lives is often tied to precise locations; their demise is a double ripping asunder. That notion dawned on me recently as my wife and I sold our home in San Antonio; she and her siblings sold their childhood abode in Kensington, California; my sisters and I sold our late mother's house on Chappaquiddick and with our paternal cousins then disposed of a number of small parcels in Chattanooga, the unbuildable remainders of a subdivision our great-great-grandfather had platted in the 1890s. "Now with the signing over of the St. Elmo property we have disposed of all our past places," I emailed my siblings. "A strange felling." They caught my Freudian slip of a typo and its unconscious elegy to those once vital relationships—spatial and familial—that have been felled.

How do we make sense of the fluctuating emotions that register in our lives with the sensitivity of seismograph's needle? How do we incorporate them into our remembrance of things past and present? And when we put pen to paper or fingers to keyboard to capture these memories, what place does "place" have in their crafting? The essays that follow speak to some of these questions, queries that allow us to seek the threads that bind us to our home-scapes, to our environs, at once historic and ephemeral, material, emotive, and recollected. This exercise, Yi-Fu Tuan would argue, is an essential act of self-discovery. If we "think of space as that which allows movement then place is pause; each pause in movement makes it possible for a location to be transformed into place."[6] His is another way of saying that our psychological yearnings, genealogical connections, imaginative play, and political desires can help us settle into place. They have helped me.

ONE Alamo City

White Gold

Some post–Civil War San Antonians loved to hate George Washington Brackenridge. His critics, in government and out, dubbed him "the Monopolist," and they had a point. Brackenridge was publisher of the *San Antonio Daily Express,* president of the San Antonio National Bank, a key investor in the local gasworks company, and the leading director of the community's privately held water works. It did not help his reputation that the man who was the city's chief creditor could seem imperious, blunt, and dismissive of those who opposed him.

Yet his financial clout, entrepreneurial instincts, and commitment to public service were essential to a city whose budget was forever in disarray and whose frugal citizenry refused to tax themselves to build vital public infrastructure, including sewers, fire hydrants, streets and sidewalks, and a potable water distribution system. The community may have chafed at its dependence on Brackenridge, but without his capital and the technological innovations his funds engendered, it would have been a far poorer place.

No one was more concerned about the need for a

healthy flow of pure water than Brackenridge. He knew, as did the physicians who had his ear, that a modern water distribution system was a necessity for a town long wracked by cholera, tuberculosis, and other infectious diseases. When he gained control of the local water company in 1883, Brackenridge donated to it hundreds of acres he owned surrounding the headwaters of the San Antonio River; his company captured and distributed its flow from this property. By 1885 he had invested in powerful pumps to pull water from the river and push it through miles of pipes serving residential and commercial customers.

Brackenridge drew on his ample resources again in the early 1890s to search out new water sources. Dr. Ferdinand Herff, a physician and shareholder in the water works, convinced Brackenridge that if the San Antonio River watershed contained an artesian structure, the pure water from this underground source would inoculate the citizenry from diseases. On his advice, Brackenridge financed a 3,000-foot test well, drilled near the *ojo de aqua*, at the river's source; it came up empty. Undeterred, in 1893 he drilled another well in the downtown core and struck pay dirt. "The fresh water supply of San Antonio is apparently unlimited," the *Daily Express* crowed. "It has been increased three million gallons for each twenty-four hours by a splendid strike . . . on the property of Colonel George Brackenridge on Market Street." That success led to others, and over the next decade new wells were bored and steam-driven pumps were purchased to flush water along what one historian has called the "mushrooming tendrils" of the Water Works Company's pipelines. This innovation ensured the year-round availability of clean, cheap water in a city long accustomed to polluted waterways and episodic drought.

Readily tapped and distributed, the water supplies nourished San Antonio's rapid population growth. In the mid-1890s the city contained approximately 40,000 people, and by 1925, when the city scraped up $7 million to buy the private water works and rename it the City Water Board, San Antonio boasted nearly 200,000 inhabitants, making it the largest city in Texas. Through the human application of technology, nature could be a handmaiden to a new, more modern city.

Or so it seemed at the time. But San Antonio's boisterous growth came with a series of environmental costs. Accelerated pumping lowered the level of the Edwards Aquifer, diminishing the natural flow from the community's many springs, reducing the volume of the San Antonio River and San Pedro Creek, and elevating anxieties about the long-term stability of local water supplies. Among those most anxious was George Brackenridge; because he lived at the river's headwaters and took a paternal interest in its welfare, he was among the first to recognize the implications of its decreased flow. The water works' artesian wells, when combined with a withering drought in the 1890s, had serious repercussions, he advised a friend: "I have seen this bold, bubbling, laughing river dwindle and fade away." This decline was so worrisome, Brackenridge averred, that he might leave San Antonio. "This river is my child and it is dying and I cannot stay here to see its last gasps."

He did not leave, and the river did not dry up, but Brackenridge's lament, uttered as it was by the man most responsible for the city's promising and problematic new water regime, reverberates still.

City Brew

Prohibition ended in Texas on September 15, 1933, and as the hour and minute hands swept in unison toward 12 a.m., San Antonio began to party. More than 6,000 citizens of good cheer milled around the front gates of the Pearl Brewery to revel in their impending good fortune; the civic and political elite jostled for a place with brewery workers and simple beer lovers. "At exactly 60 seconds after midnight," the *San Antonio Light* reported, "the steam whistle at the San Antonio Brewing Company's plant raised its voice in a shrill jubilant scream of delight." As the sky-splitting blast faded, the gates swung open and a cavalcade of horn-blaring, red-and-yellow brewery trucks rolled down Pearl Parkway. Heading toward the city's retail distributors, the drivers first had to maneuver through the boisterous throng; another 100 or more trucks followed, fanning out to El Paso, Wichita Falls, Dallas, and Houston. At the same time, an electric tram engine pushed thirty bright yellow freight cars, each packed with 1,000 cases of Pearl Beer, out of the brewery yards to hook up with a "panting locomotive on a Southern Pacific spur line." Truck and train carried the hopes of the plant's owners and thirsty consumers. "As if by magic," the *Light* noted, "the beer industry, one of the most lucrative in San Antonio, sprang back to life after its governmental 'execution' 14 years ago."[7]

Though the revelry continued into the wee hours, a more sober note was struck at the start of the business day. The brewery's personnel office was swamped with an estimated 700 applicants, only 150 of whom found immediate employment. That unsettling reminder of the Great Depression's grip on the economy was momentarily set aside in anticipation of the brewery's potential to give a much-needed shot in the arm to the region's sluggish commercial activity. After recounting the night's carnival-like atmosphere, the newspapers tabulated the economic benefits of legalized beer. In addition to Pearl, the Lone Star Brewery would soon reopen, and a third plant was in the planning stages; when the three were in full operation, the *San Antonio Express* bragged, the city would become the "leading brewery center of the southwest."[8]

New work would flow, too. There would be temporary construction jobs and permanent labor for truck drivers and brewery workers, and retail outlets—bars, beer gardens, and liquor stores—needed staffing. The city stood to gain as well. It had already seen a boom in its tax receipts; 415 permits to sell beer had been issued, netting more than $3,000 for the local treasury. Even the newspapers would profit. The *San Antonio Light* ran twenty-two pages of beer and brewery advertisements on September

15, a short-term boost that, when combined with space sold in subsequent issues, provided a modest uptick in the daily paper's revenues.[9]

It had been on similar economic grounds that Texas breweries pressed the state legislature to take advantage of the U.S. Congress decision in February 1933 to repeal the Eighteenth Amendment. To amend the constitution and legalize the production, distribution, sale, and transportation of beer and wine, two-thirds of the states would have to vote in support of the Twenty-First Amendment. Within weeks Michigan became the first state to so act, and the rush was on. To ensure that Texas followed suit, Pearl Brewery general manager B. B. McGimsey sent a letter in March, under the banner headline "What, No Beer?" to the company's distributors; he warned that they were in trouble should the state continue to lag behind its neighbors. "If Texas does not legalize beer and collect this revenue which is so badly needed, [it] will be swamped with illegally manufactured beer which will flow over our borders from Mexico and the surrounding states."[10]

McGimsey's concern for tax revenues was not entirely altruistic. He recognized that there was good money to be made in forcing, and winning, the vote in the legislature. Confirmation of this likely windfall came from reports he received from Anthony Gevers, a Pearl representative he had sent north to gauge the impact of Prohibition's demise in Illinois and Missouri, where beer was already legal. The bars were jammed, Gevers wrote, and patrons "stood there four deep, both men and women—all the tables were occupied—everybody was happy with his or her beer." That convivial moment was a revelation. "Believe me," he confided to McGimsey, "beer is the 'missing link.' " If that was the case, McGimsey replied happily, then the Pearl plant, "our old elephant," would become once more a "precious asset. It seems that the only outstanding money making business in the United States is the brewery business."[11]

Given San Antonio's historically weak industrial base, it is understandable why hopes were riding as much on the work Pearl would generate as on the beer it would brew. The Roaring Twenties had brought little energy to the city's commercial fortunes, which revolved around the same features that had driven its development since the mid-nineteenth century. Functioning as a service center for its ranch and agricultural hinterlands in South Texas and the Hill Country, San Antonio also drew heavily on federal outlays for its three major military bases (Fort Sam Houston and Kelly and Randolph Fields) and nearby installations. Because the manufacturing sector was minimal, it had long been dominated by breweries, Pearl chief among them. But Prohibition had drained them of their local significance. Through the first years of the Great Depression, the only establishments hiring workers were pecan shellers and garment sweatshops, a situation that speaks volumes about the desperate plight of the community's workforce. The opportunity to produce beer again would be significant, noted contemporary *Express* reporter Fred Mosebach: "There is no institution in San Antonio which has provided more employment than the breweries." Many felt the same way. When San Antonians voted 8 to 1 in favor of selling alcohol within county limits, they cast their vote with the expectation that Pearl's reopening would help rebuild a faltering urban economy.[12]

The Pearl Brewery had played a similar role at its founding half a century earlier. Its impact on the city's late-nineteenth-century economy had been profound, generating new jobs, spinning off auxiliary work, and fostering a strong unionized labor force. Its location also had a decided influence on the spatial design of the emerging metropolis, in particular the establishment of new neighborhoods, clustered around the plant, that housed management and labor. On many levels the Pearl Brewery had once been—and would remain for some time—San Antonio's home brew.

Nineteenth-century San Antonians had such a love affair with beer that the few small breweries operating in the decades after the Civil War could not slake their thirsts. To meet the escalating demand of 1882, for example, an estimated 400 boxcar loads of beer rolled by rail into town. The first significant manufactory, the City Brewery, was established the next year and was producing more than 150 barrels and 120 dozen bottles a day by 1884. There was a plethora of places to quaff the local brew, too. In 1887, with an approximate population of 25,000, San

Antonio boasted 169 beer-only bars and another 44 that sold beer and whiskey.[13]

These were not figures to gladden a Prohibitionist's heart, and the city was a favorite stopping point for those ardent reformers eager to do battle with the demonic liquor trade. Frances Willard of the Woman's Christian Temperance Union blew through town in 1882 and made nary a dent in the community's alcohol consumption. Twenty years later the Rev. J. K. Wooten lambasted the city's boozy reputation as he stood against the backdrop of the Alamo, perhaps seeking solace in that symbolic site of embattled liberty. He was no more effective than any of his predecessors. Drawing from the prophetic text of Habakkuk—"Woe to them that give drink to their neighbors, and thou that holdest thy bottle to their lips"—Wooten had barely started to orate when the police asked him move. He refused and was taken into custody. His shaken wife queried reporters, "What kind of city is this anyway?"[14]

The scribes might have told her that in San Antonio beer paid the bills. That would not have been strictly true, but by 1916, just two years before Prohibitionists succeeded in enacting the national constitutional amendment that shut off the tap, there were six major breweries in town; one of them, the Pearl Brewing Company, was the state's largest. They pumped hundreds of thousands of dollars into the regional economy and provided work for thousands of employees, many of whom labored for wages set at union scale; and they were among the city's first and most enduring sites of industrial production. Without beer, San Antonio would have been even more impoverished than it was.

The city might also have been culturally poorer. Much of San Antonio's social life in the post–Civil War era seemed to revolve around the world that beer made, or more precisely that which the city's large German population helped construct. Through their contributions to the community—among them, social clubs, schools and other educational facilities, newspapers, and bathhouses lining the San Antonio River—Germans sought to reproduce the intellectual energy and open-air ambience of European cities and the continental commitment to civic leisure. The Little Rhein district, a German settlement between the San Antonio River and Alamo and Commerce Streets, contained another of the urbane amenities Germans introduced to South Texas—a dozen or more "concert gardens" that reached their height of popularity between 1885 and 1900. Among the most celebrated was Scholz Palm Garden, an astonishing establishment that spanned the block between Alamo and Losoya Streets; its "three glassed-in stories, with balconies and a horseshoe bar," welcomed pedestrians inside. This "palatial beer garden," landscaped with lush tropical plants, was filled with the stirring sounds of beloved classical music that Carl Beck's house band played each night. "An orchestra hidden behind the palms played tunes by Beethoven, Meyerbeer und Wagner, staccatoed with the clinking of steins and the clicking of billiard balls," remembered Frank Bushick in his *Glamorous Days* (1934). "It was a strictly family garden and whole German families would sit around and sip their beer and enjoy themselves after the manner of life in 'the old country,' much to the amazement of provincials not accustomed to San Antonio's cosmopolitanism."[15]

Contemporary accounts were just as seductive. In 1886 the *San Antonio Light* rhapsodized over these many concert gardens' allure: "Merchants could go at noonday, eat lunch and drink enough beer to open their hearts and raise their clerks' wages. Dignified matrons not above the amber joys of honest Gambrinus could drop in quietly and refresh. Timid maidens could avail themselves of the ice facilities, and tired newspaper men drown for a while all the memory of the utter cussedness of rustling for items when there isn't one this side of the Rio Grande." The *San Antonio Herald* was only slightly more restrained in its appreciation of the imported beverage that seemed to have altered local customs in such positive ways: "In many respects we may Americanize the Germans—but in one respect the Germans are rapidly Germanizing the Americans—the Americans are acquiring a taste for the national beverage of the Germans—beer." By the work it generated and the pleasure it gave, beer had become part of the city's lifeblood.[16]

To make beer requires water, and lots of it. So it is no surprise that late-nineteenth-century investors in the City

Brewery placed their facility north of town on the eastern bank of the San Antonio River. These entrepreneurs knew they were building on preexisting development. Just blocks to the south lay a clutch of water-bottling plants and the Lone Star Brewery; the upstream waters were evidently plentiful and clean. What they may not have known was that their decision was consistent with the choices others had long made in what would become south-central Texas, highlighting how crucial the river had been to inhabitants of the region over time.

Everyone who has chosen to inhabit—temporarily or permanently—the San Antonio River valley has had to take into account the flow of water across and underneath this semiarid landscape. Whether Tonkawa or Spanish, Mexican or Anglo, each community has had to frame its social structure and economic activity around its capacity to draw on available water supplies. Late-nineteenth-century German brewmasters were no exception.

For them, the question of water supplies was particularly pressing as a result of the city's first railroad, the Galveston, Harrisburg, and San Antonio, arriving in 1877. Within a decade this east-west line was joined by one running north-south, from Chicago to Mexico, and another running northwest-southeast, from Kerrville, Texas, to Port Aransas on the Gulf Coast. These rails gave San Antonio, a once isolated frontier settlement, a series of connections to national and international marketplaces that sparked a surge in population. In 1880 more than 20,000 lived in the city, by 1890 they were more than 37,000, and in another decade they topped 53,000. Once streetcar lines were laid down, residents moved across the urban landscape with greater speed and efficiency, and housing districts and commercial activity spread out, far outstripping the capacity of the local water delivery system.

Not until the mid-1890s did a radical alteration in water supply and delivery emerge to transform the historic relationship between the people and the land. Its major promoter was George W. Brackenridge, the city's leading banker and chief shareholder of the privately owned Water Works Company. By 1885 Brackenridge had invested in pumps to suck water out of the river and push it through what grew to be 100 miles of pipeline serving residential and commercial users. His ample financial resources underwrote a series of wells drilled into an artesian plain; its discovery offered year-round availability of clean and cheap water.[17]

The realization of the Edwards Aquifer's hydrogeology, fused with importation of new pumping technologies brought about by the railroad, made it possible for entrepreneurs to found City Brewery. Yet the brewery's now easy access to water and its initial boom in sales were no substitute for good financial management, and in 1887 it went bankrupt. The San Antonio Brewing Association (SABA), which in time became known as the Pearl Brewing Company, assumed its debts and its physical plant. Like its failed predecessor, the new company—headed by

Otto Koehler and other local investors—was well aware that San Antonio's new high-speed rail connections were crucial for transporting the resources to brew, package, and distribute its product. The railroads' networks and the growing market capacity they generated also enabled the company to attract critical venture capital. Hundreds of thousands of dollars were required to salvage what remained of SABA's predecessor, and a generous infusion of capital was essential to the annual purchase of tons of high-quality barley (harvested in Missouri and Canada) and hops (grown in Washington and New York). Water was also critical, of course, but along with its liquid form, readily available from two prodigious onsite wells, the brewery consumed vast quantities of ice; its demand was so great that ice manufactory became a growing component of the economy. Beer and the resources consumed in its production generated new commercial opportunities and employment for the mushrooming metropolis.

The brewery's impact on the urban core was equally vital. Its siting close to the meandering San Antonio River was determined as much by its need for a clean flow of water as for open space in which to build and, ultimately, expand its operations. One important ramification of this location was that the eight-acre facility accelerated the spread of the city's built landscape.

Comparing data from Sanborn Insurance Maps between 1897 and 1954 with evidence of the construction of new streets, commercial buildings, and residential housing drawn from the San Antonio City Directories identifies the evolution of the surrounding neighborhoods. By focusing on specific blocks radiating immediately north, east, south, and west of the brewery, we can develop a preliminary though still rich appreciation of SABA's complicated role in transforming the community's social ecology and urban design.[18]

As the 1886 "Bird's Eye View of San Antonio" indicates, the City Brewery was set in an undeveloped stretch along the river, with no other buildings identified on an open space on the map. Sixteen years later, commencing with the 1904 Sanborn maps, the neighborhoods to the brewery's east and south had begun to develop while those to the west and north were not yet platted. By 1912 the area was mapped in all directions, albeit with varied intensity, a reflection of the population's outward thrust and the new dynamic of life on the city's expanding edge of development.

One factor in this pattern was workers' close proximity to their jobs, determined by their need to walk to their employment. An instance of this emerges in the housing of the extended Schneider family. Joe lived west of Pearl; Charles, who rented 121 Oleander (east of Pearl), worked at the Pearl Brewery. Laboring alongside him was another Joseph Schneider, who rented at 1448 Avenue B, and John M. Schneider, who rented 1507 Avenue B. Close by was John Schneider Sr., who owned and operated a grocery and saloon at 1562 Avenue B, which doubled as his residence. Assuming these Schneiders are all related, theirs is a telling example of the brewery's impact on the surrounding community: a father ran a grocery store and saloon while the sons and/or relatives worked to produce the beer he served, allowing the family to live and labor within blocks of one other.

Many members of the Koehler family served in managerial capacities at the brewery and on its board of directors. John J. Stevens was a manager for the brewery, and Andrew Stevens was its private secretary. Until 1905 or so they, like many other directors, lived nearby. Vice President Otto Wahrmund's home was located at 209 River Avenue. In 1897 Gus and John Peter were brewers for Pearl (but by 1912 they had left to open Peter Brothers Brewing, just north of their former employers). In 1905 Robert Schulz worked at the brewery alongside Adolph and Otto Schulz, an engineer and mechanic respectively. Charles and Christian Weber lived at 306 Erie Avenue and worked at the brewery (and one or both played for the Pearl Brewery baseball team). The surrounding neighborhoods were filled with similar close connections and more modest (or tentative) linkages. In 1897 Charles Rische ran the family-owned water-bottling plant near Avenue B and Grand; by 1905 his brothers Edward and Ulrich Rische managed it while he clerked at Dullnig Grocer Co. on North Alamo Street; one of his colleagues, Charles Renz, had a relative, Jake Renz, who worked at the City Brewery. Overlaps like these, in which workers, middle management, and some of the owners interacted on their way to, from, and during work, are a vivid reminder of the close-knit community that grew up around the brewery.

By 1954 the possibility of a cozy relationship between workplace, home life, and the public spaces that lay between had largely disappeared. Inhabitants of the surrounding streets were much less likely to be employed at Pearl or Lone Star breweries, as the declining cost of automobiles made it possible for postwar laborers to buy cars and commute in from newer, more distant subdivisions. Even so, the labor force at Pearl—regardless of where in San Antonio they lived—maintained high morale and a close camaraderie. "I couldn't wait to go to work," distribution manager Ron Inselmann recalled, because "it was so much fun."[19]

For all the nostalgia Inselmann's words convey, they reflect an important truth. From the beginning, those who worked at Pearl were a cohesive group. Two forces shaped this cohesiveness—ethnic identity and union membership. Beer, its production and consumption, had been imported into South Texas by successive waves of German migrants beginning in the 1840s. The city's pre–Civil War breweries, however insignificant, emerged as German-American entrepreneurs crafted small batches of the brews they and their fellow immigrants yearned to drink. By the late nineteenth century the German influence remained strong; those who underwrote, managed, or labored at the larger breweries, including Lone Star, the

City Brewery, and its successor, SABA, were invariably members of the German community.[20]

The extended Koehler family, who invested in and managed Pearl for nearly a century, were important players in this cultural connection, yet they were not alone. A cursory glance at the employee list at the Pearl plant suggests how Germanic its workforce was: men whose last names were Seidel, Staudt, Bernhardt, Hirt, Boehm, Hummel, and Krueger were among those who reaped a livelihood from their ability to transform large quantities of water, yeast, and grain into Pearl's much-touted brands, Würzburger and Texas Pride.

The sense of shared ethnicity between management and labor was no doubt reinforced through verbal communication—German was the lingua franca at Pearl until the early twentieth century (although the company's paperwork was in English). According to the 1894–96 minutes of Local 112 of the United Brewery Workers of America, because its meetings were conducted in German, the union refused membership to those who could not speak the language, a coercive measure that also underscored the Pearl workforce's linguistic identity.[21]

That its laborers were union men—among the city's first—was a source of pride and a signal of solidarity. Through their locals these men secured a closed-shop contract, determined what skills were required on the floor, and set initiation fees at a steep $5, all of which were designed to restrict employment, boost wages, and challenge management's control over the hiring and firing of workers. Among those excluded were African Americans, who, under union guidelines, could only be hired as boiler-room firemen, that most backbreaking of labors. As Judith Berg Sobré has noted, the tensions within the city's working class, and between it and management, were mirrored in the city's annual Labor Day parades. Workers rode floats or marched in formation, wearing regulation black pants, white shirts, suspenders and ties, and straw hats, their union badges on their chests. But union members were required to attend or pay a $1 fine; nonunion workers, such as the black laborers at Pearl, did not march.[22]

Segregation notwithstanding, the unionized brewery workers achieved considerable success. They secured the eight-hour day by 1901, well ahead of others in San Antonio and the nation. Their wages rose more rapidly; brewery wagon driver earnings increased from $9 a week in the late 1890s to $13 in 1901, and by 1919 they were making $20.25. More important still, workers were successful in getting management to place the "union label" on their product, meaning that Pearl—and many other breweries throughout the state—would purchase only union-made products. The payoff came when nonunion breweries in and out of Texas periodically tried to flood the market with lower-cost product. Unions worked closely with distributors to shut off the flow of "scab beer" in the Lone Star State, which not incidentally limited competitive pricing pressures on unionized plants like Pearl.[23]

Management was well aware of unionization's important role in creating a stable and productive workforce. It did its part to reinforce this outcome through Otto Koehler's strategic accommodation to arbitration and by the usual concessions during bargaining—wage increases, shorter workdays, better conditions. But Koehler and his successors also supported more informal, conciliatory gestures, including a lavish annual company picnic and a company baseball team that played in city leagues. It distributed turkeys to employees before Thanksgiving and, in advance of Christmas, handed out two cases of beer and a ham to every worker. And it maintained an employee bar. Ownership continued these traditions well into the late twentieth century and, in so doing, helped maintain and reinvigorate social bonds within the brewery. Although the Pearl Brewery may not have been the happy family its corporate paternalism was designed to encourage, neither was it crippled by the labor disputes that troubled other contemporary industries.[24]

Koehler's successors were neither so nimble nor so lucky. The American brewery industry began to change markedly in the 1950s, responding to shifts in consumer demand and purchasing patterns. Beer was increasingly sold in bottles and cans off shelves at the new supermarkets catering to those moving out to the suburbs, not from kegs tapped in bars in older urban neighborhoods. To take advantage of this national trend, large brewers began to swallow up regional competitors to reduce competition, increase market share, and invest advertising dollars in the new communication tool, television. Gregarious, bar-hopping salesmen, once the central way Pearl and other breweries had maintained their customer base, were becoming a thing of the past.[25]

B.B. McGimsey, who had been instrumental in Pearl's successful reemergence after Prohibition, understood that the business was changing, and in the 1950s he precipitated a shareholders fight to force the sale of Pearl to the Pabst Brewing Company of Milwaukee. The Koehler family resisted, vetoing the merger. But when the second Otto Koehler died in 1969, the family conceded that Pearl was no match for major breweries with greater financial resources and stronger brand recognition. They sold their holdings to the Southdown Corporation in 1970.[26]

"It is in many ways a testament to the family ethos of the business," journalist Elaine Wolff has observed, "that [the Pearl Brewery] survived as long as it did with its culture relatively intact." The enduring quality of the Koehlers' multigenerational commitment to creating an amicable workplace is apparent in the affectionate reminiscences of Pearl's employees. When he returned from World War II, Alvin Marmor, who began working at the Pearl bottling shop in 1933, "was tickled to death" to pick up where he left off following demobilization in 1945. Ron Inselmann spoke for many when he recalled what made the brewery so special: "It was big business but it wasn't big business. Everyone's door was open to you." If given the chance, he would have continued on, after Pearl closed in 2001. "If it was still open," he said, "I'd still be working there."

Like San Antonio's wild celebrations that erupted when the Pearl Brewery reopened at the end of Prohibition in 1933, Inselmann's affective ties to the brewery underscore the impact this single plant had on workers' lives and livelihoods, the neighborhoods they inhabited, and, more broadly, the indelible link between San Antonio's natural and built environments, its urban marketplace, and its cultural identity.[27]

Organizing for War

"Beat Back the Hun with Liberty Bonds," illustrator Frederick Strothmann's contribution to a World War I fund-raising scheme, is full of menace and dread. Against a smoke-filled, yellow-stained sky, a German soldier looms above a ruined Belgian city, its battered remains the reflection of a lethal onslaught; the soldier's bayonet drips with bright-red blood, as do his fingers. This terrorizing image, set in sharp contrast to the soldier's dull gray uniform and helmet and an equally blank, gray face, is magnified when you realize that this murderous figure is gazing straight at you.

It gave me chills the first time I saw Strothmann's poster, a gift from my father, who had received it from his father. Its real value, however, lies not in this genealogical lineage but in its unnerving capacity to jolt. Every year I am reminded of the poster's potent force when, partway through a lecture on propaganda's role in shaping American responses to the war, I carry it around my classroom so that my students can see what I am trying to describe in words. They recoil as if hammered by the kick of a Springfield rifle, standard issue to the American Expeditionary Forces in Europe. Stunned by its continuing capacity to evoke fear and disturbed by its dehumanizing brutality, they come away with a better sense of some of the impulses that framed that earlier generation's declaration of war.

These students at Trinity University, located just north of downtown San Antonio, are also startled to learn how important the city and its hinterlands were to the formulation of the war effort. Although thousands of miles away from the ghastly trenches in eastern France, buffered by the Atlantic Ocean and much of the North American continent, and therefore immune to the vicissitudes of a devastating conflict that claimed more than 20 million lives, the Alamo City and its environs helped place the nation on an aggressive wartime footing.

This claim about San Antonio's manifold contributions to the U.S. decision to join arms with the Triple Entente (Britain, France, and Russia) in its fight against the Central Powers (Germany, the Austro-Hungarian Empire, and Italy) is rarely made in analyses of the tumultuous geopolitical situation that let loose the guns of August 1914. Understandably so—far more compelling is the focus on the late June 1914 assassination of Archduke Franz Ferdinand, heir to the Austrian imperial throne, which triggered a series of linked declarations of war between European nations great and small. Two months later the continent was caught in a brutal, protracted struggle of its own making.

The official American response was to remain neutral, a strategy President Woodrow Wilson affirmed over the next two and a half years. Although his position dovetailed with public opinion, it became increasingly difficult to maintain as the German and British navies sought to use their competitive advantages—submarines and ships, respectively—to destroy the other's capacity to fight. In the middle of this intensifying battle was the American merchant fleet. As it plied the North Atlantic, carrying food, fuel, and weapons to the belligerents, the vast bulk of which ended up in French or British ports, the Germans began to target this vital supply line; their torpedoes began to take a toll on the flow of cargo.

By early 1917 their attacks on passenger liners (notably the 1915 sinking of the *Lusitania*, which killed 1,198, including 128 from the United States) had intensified American sympathies for the Triple Entente's cause. With pressure mounting, on April 2 the president called Congress into "extraordinary session because there are serious, very serious, choices of policy to be made, and made immediately, which it was neither right nor constitutionally permissible that I should assume the responsibility of making."

In his request Wilson laid out this carefully reasoned legal brief: because the German use of submarines violated international law; because their destructive force was aimed at the "wanton and wholesale destruction of the lives of non-combatants . . . engaged in pursuits which have always, even in the darkest periods of modern history, been deemed innocent and legitimate," it was manifestly clear that neutrality was "no longer feasible or desirable." Under these conditions armed confrontation was the only answer. "The world must be made safe for democracy."

Wilson's words hit home in San Antonio. As one of the city's dailies, the *Light,* observed, "The war between the United States and Germany which all farsighted people have long beheld as a possibility, and which all right-minded people have conscientiously endeavored to avoid, has come." It came, the newspaper asserted, because "our good intentions have been misconstrued, our patience has been abused, our forbearance has been outraged." It was left to Americans to "prove to the world that, although a peace-loving people, we are still the children of those who . . . have never failed . . . to finish in their own way those things to which they have set their hand."

The newspaper's readers no doubt thrilled to these patriotic sentiments, for San Antonio had long been at the center of and defined by armed conflict. Late-seventeenth-century Spanish explorers and missionaries had established it as a fortified outpost on the northern frontier of New Spain. The town served the same function after Mexico overthrew its Spanish overlords and held a similarly strategic position for the Texas Republic (1836–45)—remember the Alamo?—as for the United States, which took possession of the region in the late 1840s. That San Antonio was precariously located on the border of these various expanding empires was not by happenstance. Its physical environment shaped its political fortunes. Spanish planners had sited it at the break point between the Great Plains to its north and the coastal grasslands that rolled south, setting it beneath the folds of the rugged Edwards escarpment. Bubbling up from this rocky abutment was a series of springs whose crystalline flow fed the San Antonio River and San Pedro Creek, and that caught the eye of Spanish commanders. Here was open, defensible, and well-watered ground, a perfect location for a garrison town.

Every subsequent military power whose flag flew over the Alamo and Main Plaza reached the same conclusion. Cementing this enduring relationship was the late-nineteenth-century construction of sprawling Fort Sam Houston on an elevated stretch of land immediately northeast of the central core, an installation that for a time would hold the distinction of being the nation's largest. During the Spanish-American War of 1898, for example, Fort Sam trained military units, including Teddy Roosevelt's Rough Riders, a capacity that would expand with the creation of American hegemony across the Caribbean. Between 1898 and 1917 the United States stationed troops in Puerto Rico, the Panama Canal Zone, and Cuba; established formal protectorates over Haiti, the Dominican Republic, and Panama; and periodically occupied each of these countries, as well as Nicaragua, Honduras, and Mexico. A large contingent of these servicemen first earned their stripes in Fort Sam's classrooms, at its firing ranges, and on its parade grounds.

These overseas engagements proved formative in another sense. They were rigorous testing grounds for American armed forces and their equipment in the run-up to World War I. Most immediate of these was the 1916 Mexican Expedition, also known as the Punitive Expedition. The strike force, commanded by Gen. John "Blackjack" Pershing then headquartered at Fort Sam Houston, was sent to the U.S.-Mexico border to prevent Mexican guerillas from attacking across the Rio Grande. Most famously, it punched deep into northern Mexico in a futile attempt to capture revolutionary Pancho Villa, whose troops had raided Columbus, New Mexico, in early March 1916. More than 150,000 members of the National Guard were called up as part of the action, a large number of whom were shipped to San Antonio to receive basic training at Camp Wilson, practice gunnery at Leon Springs Military Reservation, and learn tactical maneuvers at Camp Bullis. Although they failed to take Villa into custody and were mustered out of federal service in late 1916, these newly hardened soldiers and officers learned valuable lessons that they would put to good use when they were remobilized months later on April 6, 1917, to sustain Congress's declaration of war on Germany.

The long-awaited resolution, cheered across the country, was a boon to San Antonio. Its extensive history as a military service center and staging ground, a mission the Army had expanded with the 1915 construction of its first aviation facility on the grounds of Fort Sam Houston—called Dodd Field—ensured a remarkable influx of trainees for the duration of the war and a steady infusion of federal dollars into communal coffers.

Bases dominated the local map. Camp Wilson was renamed Camp Travis and overnight became a self-functioning city replete with 1,400 temporary buildings in which more than 100,000 soldiers (approximately 10 percent of all Americans who served in Europe during the war) were housed, fed, taught, and trained.

Dodd Field could not accommodate the Army's escalating demands for flight-ready pilots and ground crews, so the Department of War built Kelly Field on 700 leased acres of ranchlands south of the city's downtown core; within a year it added another 1,800 acres. Environmental factors were crucial to this decision, as the open, flat terrain dovetailed with the region's benign weather and clear skies, making South Texas a perfect site for the vast training facility. This rapid expansion was also impelled by Kelly's mind-boggling training schedule. On Christmas Day 1917, for example, more than 39,000 men were stationed at the field; the next month, despite shipping out 15,000 cadets, another 47,774 recruits had arrived. Cumulatively, "Kelly soldiers organized approximately 250,000 men into aero squadrons during the hectic months of 1917 and 1918. The Enlisted Mechanics Training Department turned out an average of 2,000 mechanics and chauffeurs a month. Most of the American-trained World War I aviators learned to fly at this field, with 1,459 pilots and 398 flying instructors graduating from Kelly schools during the course of the war." Among these was a local boy made good: Edgar Tobin, a member of Capt. Eddie Rickenbacker's "Hat in the Ring" air squadron and a highly decorated ace.

Kelly's boom was such that another airbase, Gosport (later Brooks) Field, had to be carved out of the city's south side; on its 1,300 flat acres, pilots were trained to fly balloons and airships. Civilians pitched in as well. Marjorie Stinson, who had trained fliers before the war at Fort Sam Houston with her daredevil siblings Katherine and Edgar, continued doing so once hostilities began, operating out of Stinson Field, the city's municipal airport. Soldiers marched; planes whizzed overhead; rifle fire crackled; and howitzers thundered. However far removed it was from the major battlefields at Verdun, Ypres, and the Somme, San Antonio must have sounded and smelled as if it was in the midst of war.

Walking its streets would have confirmed as much. Uniforms were everywhere; military vehicles crisscrossed the city between its many bases, depots, and camps; and entrepreneurs eager to cash in on wartime spending were in abundance. Some visionaries hoped the war would increase the city's industrial base—and what better item to manufacture than airplanes? Their ambitions made sense, but inept leadership, divisive politics, and the failure nationally of postwar military and civilian aircraft production to take off scuttled the idea. Yet out of this civic activism emerged a greater emphasis on military spending in San Antonio's economy, and thus a greater dependence on its appropriations. In times of conflict, the citizenry would discover, this could be a good thing; in peace, as federal funding dwindled, so would their fortunes.

That said, the population grew: 96,000 lived in the city in 1910, and at the close of the war it was home to more than 160,000. These figures do not factor in the waves of temporary residents who flooded the town for basic training or advanced instruction. Few had weathered anything like the region's withering heat or sticky humidity, its furious thunderstorms or bone-chilling "blue northers" that screamed down the plains. One pilot remembered "arriving at Kelly Field after withstanding a long, hard trip, when food had been given out 36 hours early, with great anticipation of becoming a great aviator and of making fame by bombing old Hun 'Bill's' palace," only to snap to attention when he found himself "lined up in front of a row of tents . . . feet in black mud and the wind blowing a gale, trying to obey the orders—'prepare for inspection.' "

No less bewildering were newcomers' close encounters with San Antonio's notorious red-light district, a vast warren of bars, gambling emporia, and brothels west of the central business district. It reminded an aviation cadet of what he imagined a "roaring, wild west town" would look like; the city "seemed to abound in saloons that were decorated with extremely long cow horns on the walls, and with glass enclosed, realistically preserved habitants of the state, including coiled rattlesnakes and hairy tarantulas, to greet the visitor's startled eye from many nooks and crannies." The city was an unusual place in a disorienting time.

Then hostilities suddenly ceased. The armistice was signed on November 11, 1918, seventeen months after the formal American declaration of war. The news touched off enthusiastic demonstrations from coast to coast, and ticker tape swirled. Rival papers dueled over the most punched-up headline, and San Antonio was no exception. "EMPEROR ABDICATES," the *San Antonio Light* trumpeted; "GERMANY GIVES UP," the *San Antonio Express* blared.

Back at Kelly Field, at least initially, the trainees were in no mood to celebrate. When word reached them there "was dead silence for about 30 seconds, and then the whole barracks sat up and started cursing at the floor, damning the Germans for having lain down on the job before they had a chance to get over there and prove their merit as aviators."

Their commander-in-chief knew better than to prolong the bloodletting. More than 100,000 Americans had lost their lives, another 200,000 had been injured, and the nation's wounds—social, political, and economic—went deeper still. That is why President Wilson's "Thanksgiving Address," delivered within days of the armistice, spoke only of hope: "Complete victory has brought us, not peace alone, but the confident promise of a new day as well, in which justice shall replace force and jealous intrigue among the nations." Lauding our "gallant armies" and the "righteous cause" they battled for, Wilson knew that the American Expeditionary Force had "nobly served their nation in serving mankind." Their accomplishments were a source of great joy: "We have cause for such rejoicing as revives and strengthens in us all the best traditions of our national history. A new day shines about us, in which our hearts take new courage and look forward with new hope to new and greater duties."

It was not to be. On the international front, Wilson's generous fourteen points for postwar reconstruction, which included the creation of a League of Nations to guarantee political sovereignty and territorial independence, were undercut by his European allies and his congressional opponents. The domestic arena was troubled as well by a freefalling economy, the deadly influenza pandemic, Red Scare deportations and imprisonments, a brutal crackdown on labor unions, and vicious race riots that ripped through more than twenty-five cities, including Chicago, East Saint Louis, and Tulsa. War's end brought little peace.

San Antonio experienced some of this turbulence. As its military infrastructure shut down, commercial activity collapsed, leaving the city's newfound workforce with fewer employment options. That dispiriting situation stabilized by the mid-1920s, only to worsen with the onslaught of the Great Depression, calling into question the wartime sacrifices so many residents had made—their Liberty Gardens, the natural resources conserved, the War Savings stamps purchased, and, most devastatingly, the loved ones they had lost.

These understandable concerns aside, San Antonio had directly benefited from World War I. In absorbing millions in federal largesse to train hundreds of thousands of troops, it marshaled its natural and human resources, expanded its rich military legacy, and built up a robust urban economy all while helping to "Beat Back the Hun."

Political Legend

Just hours before he delivered one of the most important speeches of the 1960 presidential campaign, an address to the Greater Houston Ministerial Association in which he argued that "because I am a Catholic, and no Catholic has ever been elected President, the real issues in this campaign have been obscured," John F. Kennedy stood in front of the Alamo.[28]

An estimated 30,000 San Antonians gathered to hear the charismatic Democratic candidate. Flanked by native son and vice-presidential running mate Lyndon Johnson and a host of local politicians, Kennedy spoke of America's future by reflecting on its past. "We honor the independence of Texas today," he affirmed at the site heralded as a cradle of that liberty, and then promised, amid heightened Cold War tensions, to recommit the nation to the advance of democracy. "The new frontier of which I speak does not consist of the things which we promise we will do for you. It consists of the things which you can do for your country, the opportunity for service, the opportunity to help this country realize its great potential, here and around the world. In the American Revolution,

Thomas Paine said, 'The cause of America is the cause of all mankind.' I think in 1960, the cause of all mankind is the cause of America." Kennedy left the podium to thunderous applause.[29]

Although his presence in San Antonio was part of a preelection swing through Texas, and his Alamo address was one of many stump speeches, the embattled setting of this particular talk played a critical rhetorical function in Kennedy's two talks later that day in Houston. At the Houston Coliseum he opened with a joke in which a Texan brags to a Bostonian "about the glories of Bowie, Travis, Crockett, and all the rest." After a while the Yankee retorts, "Haven't you heard of Paul Revere?" Of course he had, the Texan replied. "He is the man who ran for help." The audience roared with laughter, which redoubled in energy when Kennedy added a self-deprecating twist: "I am down here in Texas running for help."[30]

No such affectionate outpouring greeted him at his next stop, the Rice Hotel, where he spoke to a large gathering of Christian ministers. Many appeared convinced that, as president, Kennedy would be the Vatican's vassal. How fitting that a onetime Catholic mission, the Alamo, provided him with an opportunity to defuse Protestant anxieties about the presumed demands of his faith on his potential constitutional responsibilities. Kennedy promised to uphold the nation's revolutionary commitment to "the kind of America for which our forefathers [died] when they fled here to escape religious test oaths that denied office to members of less favored churches—when they fought for the Constitution, the Bill of Rights, the Virginia Statute of Religious Freedom—and when they fought at the shrine I visited today, the Alamo." Inside the aged mission's walls, Kennedy asserted, "side by side with Bowie and Crockett died Fuentes, and McCafferty, and Bailey, and Badillo, and Carey, but no one knows whether they were Catholics or not. For there was no religious test there."[31]

In three speeches over twelve hours, Kennedy used the Alamo's complex history and symbolic import to shore up political support and tamp down sectarian opposition. He was neither the first nor the last to take advantage of the famed battleground's affective claims, and his handling of the memories it evoked in his varied audiences, however deft, were not unique. The Alamo has long been a contested space, cultural flashpoint, and durable motif.

Which begs the question, how are we to remember the Alamo? That depends, anthropologist Holly Brear argues in *Inherit the Alamo*, on who asks the question and what answer(s) he or she desires. "Because the Alamo is the purported origin of Texas society, claiming its past is a principle means of establishing groups and individuals as being heirs to the present." This process, in which "we fight the social and political Other . . . with images and words rather than with guns" and thus "create boundaries between 'us' and 'them,' " ensures that our "historic battlefields remain our battlegrounds." An outcome that will extend across time, this tension about the site's contemporary ritual power and symbolic resonance, "despite [their] references to the past, is for the future."[32]

That's why controversy swirls around what is deemed historically significant about the building complex known as the Alamo. Its iconic stature was sealed, for many, with the March 1836 battle pitting a small band of Texians and Tejanos against the much larger Mexican Army under the command of Gen. Antonio López de Santa Anna. This struggle between insurgents seeking independence from Mexico and the Mexican nation-state, which sought to extend its central authority over the rebellious northern province of Tejas, ended on March 6; no quarter was given to the defenders.

This fatal climax has been heralded ever since. As the Alamo's official website asserts, "Although The Alamo fell on the early morning hours of March 6, 1836, the death of the Alamo Defenders has come to symbolize courage and sacrifice for the cause of Liberty." Or, as historians Randy Roberts and James S. Olson observe of the insurgents, "they achieved a certain immortality. Today, millions of people visit the place where Travis, Bowie, Crockett and others perished," and these many visitors' presence adds weight to the Alamo's claim on "history and memory, as alive today as it was in the nineteenth century."[33]

But the Alamo also contains an earlier and equally important set of meanings linked to its 1718 founding as Misión San Antonio de Valero. It was established to protect Spain's claim to what is now south-central Texas from the Lipan Apache and the French. Those who lived within its walls, missionaries and Indian converts alike, were also to spread Catholicism and boost local economic development. They did so in combination with those stationed at four other missions located south along the San Antonio River, and with the civilian population of San Antonio de Béxar, situated across the river to the west. Together this sprawling settlement maintained farms, herded livestock, and conducted trade along El Camino Real de los Tejas. In 1793, after seven decades of service, the missions were secularized and their lands distributed to their residents, a pivotal moment of cultural transformation, social maturation, and religious liberation that predates, is independent of, and thereby complicates the subsequent fixation on a bloody fight that transpired at the Alamo more than forty years later.

Once secularized, the mission evolved into a strategically significant military redoubt. For the next thirty years it served as a Spanish base for military operations, gaining a new name from a cavalry unit from Alamo de Parras in Coahuila. Revolutionaries opposed to Spain's imperial presence later captured it and launched sorties from the site. In 1821, after Mexico gained its independence, the new nation's flag flew over the building until December 1835, when another insurgency erupted in San Antonio; after fierce house-to-house fighting, Gen. Martín Perfecto de Cos surrendered the Alamo to a revolutionary force intent on liberating the province. As a start, Col. William B. Travis and a ragtag collection of volunteers began to refortify the Alamo, hoping to repel Santa Anna's Army of Operations, numbering over 4,000.

They failed. Although the insurgents lost this battle, six weeks later their comrades-in-arms would win the war. On April 21, at San Jacinto, an army under the command of Sam Houston defeated Santa Anna's troops and captured the Mexican commander-in-chief and president. As part of the terms for his release, detailed in the Treaty of Velasco that the warring parties signed on May 14, 1836, Texas gained its independence. Even before that moment, the Alamo had gained legendary status; Sam Houston's soldiers are said to have shouted "Remember the Alamo!" as they swept to victory.

And to these victors went the spoils. The legendary clash at the Alamo has had a considerable impact on San Antonio's social structure, economic development, and spatial design. Within thirty years the Spanish town had become an American city. Its population, which diminished to less than a 1,000 after 1836, rebounded to more than 3,000 in 1850, and by 1860 it had topped 8,200, the bulk of whom were Anglos and Germans. The Mexican population lost its former demographic significance and economic clout, becoming second-class citizens in an English- and German-speaking community.

Alamo Plaza reflected these changes. The walled-in enclosure was torn down, and the open space was restructured to facilitate the flow of commercial traffic, with a park platted in its center. Some stores that had surrounded Spanish colonial Main Plaza to the west migrated to the newer, more American streetscape around the Alamo. The area also became the location of the U.S. Post Office, hotels, and saloons. The Alamo itself became the headquarters of the U.S. Army's Quartermaster Corps. These alterations, anthropologist Richard Flores notes, signaled "the emergence of a new social order that revalue[d] property as capital, privatize[d] the conduct of commerce and leisure, and redefine[d] the public from a social collectivity to one based on class location and privilege."[34]

Some of this is overdrawn. Main Plaza was still the locus of the city's commercial and banking activity, and it remained the hub of its spiritual life. Still, in the late nineteenth century Alamo Plaza superseded Main in this crucial respect: efforts to preserve and heighten the particular memory of the 1836 battleground, to consecrate its embattled grounds, nourished a robust tourist economy. Central to that preservation project was the articulation of the Alamo as the site of a fight pitting Euro-American heroism against Mexican despotism, courageous whites battling cowardly browns. Its constituent (and racist) elements became the thematic thread that bound together popular literature, tourist guides, and public discourse and fused them to a once neglected building—one that, when refurbished, achieved landmark status under the management of the Daughters of the Republic of Texas.

Without the Alamo, an old joke goes, there would be no Texas, the political meaning of which anthropologist Flores expands when he concludes, "the heroic, mythic tale of the Alamo is . . . a story about the birth, not merely of Texas, but of the United States and the western frontier." While the West "as a place . . . was surely present before 1836, it is the West—as a project of modernity—that emerges full force with the cultural birth of the Alamo."

Jack Kennedy spoke to this modernizing West when he toured Texas in 1960 in search of the votes that ensured his place in the national political landscape. By coming west he also identified the increased power and cachet of this region, its centrality in the American imagination and electoral life. He may not be the last northeasterner to become president, but to date there has not been one since his election. Little wonder that many of his successors would make their own pilgrimage to the Alamo.

Buyer's Remorse

It was over in a matter of minutes. The 1960s ranch house proved no match for the bulldozer, which crashed first into its east wall, flattening its brick facade and splintering its interior framing before knocking over the remaining structure. As a cloud of dust rose over the mound of crushed sheetrock, twisted two-by-fours, and bent piping, a demolition crew swarmed over the site and within forty-eight hours scraped it clean.

There was nothing unique about this teardown in Olmos Park, a 1920s San Antonio automobile suburb. It was the second in as many months in my community, a pattern that escalated in the first decade of the twenty-first century, making our experience part of a national trend fueled by speculative development and skyrocketing urban land values. Neither is it a shock that the multistory edifice soon to occupy this lot nearly tripled the square footage of its low-slung, midcentury modern predecessor. To draw attention to this disturbing process, in February 2006 Preservation Texas released its annual list of *Texas' Most Endangered Historic Places* and for the first time warned that many of the state's "historic and architecturally most significant neighborhoods" were under assault. From the Colonial Country Club district in Fort Worth to Dallas's Old Preston Hollow and Bluffview as well as Austin's Pemberton Heights, older homes were being demolished—an irreplaceable loss of architectural integrity, of "character and charm."

The same holds true in the Alamo City, which contains the largest number of endangered first- and second-ring suburbs. The incorporated sister cities of Alamo Heights, Terrell Hills, and Olmos Park, like their older unincorporated peers, Monte Vista and Beacon Hill, exemplify the deleterious impact "boxy new mansions" are having on streetscape. Their swollen footprints are eliminating mature trees and landscape features, diminishing front- and backyard sociability, and, with the destruction of affordable homes, reducing potential economic and demographic diversity.

This homogeneous impulse—more evident in some sectors than others—is the mirror image of the mid-twentieth-century white flight that had such damaging consequences for American cities. After World War II, as members of the middle class drove out of the urban core to its expanding periphery, they abandoned streetcar suburbs and their mix of populations, creating a social ecology that starkly segregated metropolitan areas by race, ethnicity, and class. The accelerating process of mansionization, Preservation Texas and neighborhood activists fear, threatens to transplant these debilitating social divisions to inner-city neighborhoods.

That particular worry was misplaced about Alamo Heights, Olmos Park, and Terrell Hills, whose founding covenants long ago established them as white-only enclaves. Yet even these wealthy havens have been buffeted by legitimate anxieties related to a radical shift in aesthetics. The teardown mania and steroidal remodeling craze in Alamo Heights, for example, have so badly disfigured its once cozy "cottage district" that it is no longer worthy of its humble moniker.

That said, preserving the visual legacy of built landscapes should not mean freezing the past, making it static. History's powerful hold on our imaginations should not deflect us from recognizing that human habitats—like their natural counterparts—are dynamic and must remain so. If by a lockstep preservationism we shut down their capacity for renewal, for a vital interplay across time, we will stunt the regenerative possibilities of a rich and evolving communal life.

Yet a clear-cut terrain can have the same destructive impact. That's why it is essential for cities like San Antonio and its incorporated suburbs to enforce existing demolition permits and enact similar tough regulations in unprotected communities. No less critical is a sustained and rigorous education of homebuyers, developers, and real estate agents about the enduring value of building or remodeling houses that reinforce rather than degrade neighborhoods' architectural sensibilities and material vocabularies. If builders and neighbors need blueprints to draw from, they have only to look at the eclectic environments being toppled by the bulldozer's blunt force.

Danger: Work Ahead

Highways have histories. That's easy to grasp when discussing that marvel of Roman engineering, the Apian Way, or the equally remarkable 14,000 miles of Incan roads that united a far-flung Andean empire, or the Third Reich's bid for transportation immortality, the Autobahn. But what of San Antonio's major north-south highway, U.S. 281?

That query popped up as I motored north on McAllister Expressway one early January evening, running late for a Smart Growth meeting on the controversial plans of the Texas Department of Transportation (TxDOT) to widen the freeway and add toll lanes. Two things—fuel and speed—allowed me to arrive in time to protest the very highway I had cruised along, a complicity of historic dimension. For I am not the first auto-fixated San Antonian to challenge this concrete artery's existence.

In 1960 the city conducted a bond election to fund U.S. 281's initial construction, prompting the San Antonio Conservation Society to go into vocal opposition. "STOP THIEF," blared one of its flyers, which asserted that the proposed expressway would "destroy the world famous SUNKEN GARDENS" and ruin public "PICNIC GROUNDS and RECREATION AREAS." The attack helped defeat the bond, and *Life* magazine captured the moment's significance: "A new breed of engineers regards concrete, anywhere, as more esthetic than nature, and sometimes need to be put in their places by alert and stubborn conservationists—as San Antonio did."

The engineers did not stay in their place for long. The next year a new bond package swept the polls, sparking a decade-long brawl between protestors and city hall. Although the road was completed, there was a price. Sen. Ralph Yarborough (D-Texas) amended the 1966 act that created the U.S. Department of Transportation to prevent its secretary, in the senator's words, "from unleashing the bulldozer on our public parks, historic sites, wildlife refuges and recreation areas."

That amendment initiated the idea of environmental impact statements, setting in motion the passage of the 1969 National Environmental Policy Act (NEPA), whose protections figure in the most recent legal challenge to U.S. 281. In response to TxDOT's plans, in December 2005 Aquifer Guardians in Urban Areas (AGUA) and People for Efficient Transportation filed for an injunction in federal court, citing the lack of a full-scale environmental assessment; one month later the Federal Highway Administration concurred, temporarily halting construction. The past has a funny way of shaping its future.

This forty-year history of San Antonio freeway fights

is not unique. In the 1960s grassroots organizers in New Orleans and Boston used the Yarborough amendment to stymie the highway lobby; in contemporary Los Angeles, midwestern metropolises, and Sunbelt mega-cities, citizens are rebelling against automobile-powered sprawl and a Wal-Martized terrain. These struggles will continue, for activists everywhere are confronting a set of interrelated issues that are a consequence of skyrocketing urban populations. Such mushrooming growth is of special import to the American Southwest, home to six of the nation's ten largest cities, from Houston to Los Angeles to San Diego. Each faces daunting challenges in resource management and human services, water distribution, food insecurity, and educational inequities—pressures that will swell with every new inhabitant.

How are we to build more sustainable cities? Robert Gottlieb argues in *Forcing the Spring* (2005) that the answer lies in an environmental movement explicitly advocating for "nature and the poor." The intellectual origins of his argument lie in the Progressive Era. President Theodore Roosevelt and his successors set up federal regulatory agencies that tapped the energy of a powerful conservation movement focused on environment dilemmas, natural and human. Growing out of this Progressive ethos was landmark legislation that subsequently swept through Congress from the mid-1960s to the mid-1970s, including the Wilderness Act, NEPA, and the Clean Air and Water Acts. AGUA is part of a century-long effort to construct a more habitable society.

But modern-day progressives face a stiff challenge. Ever since the Reagan Revolution, Republicans have assaulted the Endangered Species Act, undercut NEPA, and diluted air- and water-quality regulations. Meanwhile the environmental left has rejected top-down, expert-driven solutions that flowed from federal environmental agencies and has severed ties with professionals at the Sierra Club and the National Resource Defense Council for their failure to attend to issues of environmental justice. The twin attacks may have opened the way for an environmentalism that is, in Gottlieb's words, more "democratic and inclusive," a transformative movement framed around "equity and social justice," but the attacks have also complicated activists' efforts to develop durable coalitions, a necessary precondition for enacting political change.

AGUA's experience is reflective of this complex situation. To hold back the hitherto unchecked suburbanization of the Hill Country and the compromising of its underground water supplies requires a broad constituency willing to tie upstream development to downstream public health issues. To appeal to other, less advantaged sectors of the city, AGUA should reembrace the strategy it and its allies employed in 2003 to delay construction of the PGA Village in Florida; then it successfully argued that a northern-tier golf course must contribute to the whole community's well-being, supporting Communities Organized for Public Service's living-wage demands, a connection that advanced each other's causes.

This inclusive strategy alone will not bring success. AGUA and other grassroots activists need the legislative savvy and financial clout that more mainstream groups marshal, the same organizations, such as the Sierra Club, that local environmentalists often demonize. Only by pairing local knowledge with national muscle will AGUA boost its chances of stopping the earthmoving equipment in their tracks.

Holy Moses!

Moses had a burning bush. San Antonio had the Helotes mulch fire. Each was a signal moment, each a call to action. Moses paid heed to that fiery bush, from which he received God's charge to lead his people to freedom. San Antonio would do well to emulate Moses's alert response to the message embedded in the abnormal. But given the community's reactions to the Helotes blaze that erupted during winter 2007, I have my doubts.

The controversy surrounding the smoky conflagration in the Alamo City's northwest sector began in February 2006, when the Texas Commission on Environmental Quality (TCEQ) grew concerned at the swelling size of the Zumwalt Company's mulch pile. Over the years it had grown to nearly ninety feet high, a football field in width, and several hundred feet long, a monster mound of tree trunks and wood chips that appeared to violate the company's state permit. But TCEQ did nothing. It responded similarly the following December when the massive pile began to smolder, thin trails of smoke rising out of its dark looming mass.

When the flames became more visible and the

smoke thicker, and when the inescapable pall made for a stomach-churning experience for those living nearby, the state agency and local governments took notice—that is, they delivered platitudes until the media turned up the heat. Finally, in a desperate effort to achieve a quick fix, TCEQ hired Oil Mop LLC to deconstruct the pile while it hosed down the burning material.

In an ideal situation that would have been the end of the story—a recalcitrant governmental agency, shown the error of its ways, moved with dispatch to restore order. The people of Helotes and far beyond would have breathed a huge, clean sigh of relief.

But this was not a perfect moment of redemption. We were looking at the burning mulch, but, unlike Moses, we ignored its significance. Consider, for example, the failure of the state body and local authorities to establish the fire's impact on air quality. Although those in the surrounding neighborhood, including O'Connor High School, provided ample evidence of respiratory problems, the scientific monitors did not register adverse data.

San Antonio's Metropolitan Health District would later admit that its instruments were in the wrong position to capture the information necessary to assess how compromised public health had become. It was unsettling that in a region facing rigorous sanctions for its nonattainment of Environmental Protection Agency air quality standards, its pollution experts had botched this rudimentary evaluation.

Much worse was TCEQ's staggering inability to appreciate the fire's threat to the Edwards Aquifer, the city's sole water source. Oil Mop's suppression strategy, which TCEQ certified, included dousing the massive mound with millions of gallons of water. But neither TCEQ nor Oil Mop took into consideration the fact that the inflamed pile was directly over the aquifer's sensitive recharge zone. As arcs of water were hosed onto the flames, the resultant carbon-stained fluid seeped into the ground, trickling through cracks and crevices into the aquifer. Within days neighboring wells sucked up gray-tinged water.

Even more aggravating was that the Edwards Aquifer Authority and the San Antonio Water System had given TCEQ advance warning of this outcome. These resource organizations' vigilance was one of the silver linings to this upsetting episode. Their actions, moreover, gave the state legislature the evidence it needed to finally understand why local agencies must be empowered to better protect the 1.5 million people dependent on the aquifer. Alas, legislators did not take the hint, and during its subsequent 2007 session it refused to grant either San Antonio or Bexar County stronger regulatory authority over development in and around the aquifer's drainage and recharge zones.

The good news was that the Helotes mulch fire was extinguished in late March 2007. The bad news was that the community could not rest easy. It could not because the fire was a direct result of the city's surging growth dating back to the post–World War II era. Beginning in the 1960s San Antonio, which then claimed more than 600,000 people, had begun to sprawl north and west, building out first to Loop 410, a fifty-mile circle around the city. Within twenty-five years the city of nearly 800,000 had pressed out to Loop 1604, an expressway 100 miles in circumference. Since then, with a population nearing 1.5 million, the city has pursued the same developmental pattern, pushing deep into the Hill Country. This rural and rolling terrain has been suburbanized, with new housing subdivisions and gated communities, shopping malls, roadways, and schools sprouting up with alacrity. To erect this built environment, the first things to be cleared away were the native trees, shrubs, and groundcover that attracted people to this rugged landscape in the first place. Truckload after truckload of plant material was hauled away to dump sites scattered across the region. The Helotes pile was one of many repositories of woody debris, some of which are also upstream of or right over the aquifer's recharge zone. The choices so many have made about where and how to live were implicated in the Helotes fire.

So they will be if another blaze erupts. If it does, the community will no longer be able to plead ignorance about the belowground consequences of aboveground actions. The Helotes mulch fire illuminated the ineluctable connections between hydrology and demography, and San Antonio's need to bring these into alignment quickly.

Absent the essential human commitment to environme
tal stewardship, San Antonians had better pray for divine intervention. The flickering flames of Helotes suggest how little time the city—or anyplace like it—may have to save itself from itself.

Repairing Eden

ture is the preservation of wildness.

erry

What is the human place in Nature? Sorting out that relationship is at the heart of contemporary debates swirling around landscape-restoration projects in the United States. This is as true for large-scale land management schemes devised to return millions of acres of our national forests to greater health as it is for the seemingly more modest ambition to reclaim a battered nineteen acres in Cathedral Park set within San Antonio's Olmos Basin. Yet the scale or size of a project is less significant than the intellectual tensions and cultural dilemmas that invariably ripple out from the initial decision to restore a damaged ecosystem. When we ask what our connection is to the landscape we have decided to rehabilitate, unstated is the nagging thought that we have no right to intervene. Nature's ends and humanity's means often appear antithetical.

There is truth to that sentiment. After all, no species is more invasive than *Homo sapiens*. Our capacity to consume resources and carve up landscapes to serve our needs—real and imagined—is astonishing, even terrifying. Although the bird's-eye view of our sprawl that airline travel offers is unique to our time, we are hardly the first generation to feel uneasy with our rapacity. We have long imagined ourselves as Man the Destroyer, the Defiler of Sacred Space, a damning self-perception reaching back to one of our most ancient stories, that of Adam and Eve. They were expelled from Eden because Adam sinned, and we continue to do wrong, compelled to live apart from the nature that sustains us.

Its explanatory power notwithstanding, this troubling vision of our debilitating sense of uprootedness misses an important point. Adam's life, as narrated in Genesis, contains a resolution for our modern environmental anxieties. Take his creation: "God formed man [Adam] from the dust of the earth [*adama*]" (Gen. 2:7). Human form was conceived out of the ground we have walked on since. An essential element in this place, we cannot relinquish that birthright or the obligation it imposes on us, an obligation we are reminded of by the task the Divine gave to this new man. Adam's work, we learn, was to "till the soil from which he was taken" (Gen. 3:23). A farmer, a steward, Adam may have been thrust out of Eden, but only by being expelled could he fulfill his critical mission—to repair his shattered soul and stitch together a broken world, to care for the terrain he would find his living in, "by the sweat of [his] brow" (Gen. 3:19).

Even the great flood, which drowned so much, could not sweep away humanity's primal link to and intimate relationship with the land. No sooner had Noah departed from his beached ark than he performed a vital act. This "tiller of the soil" got down on his knees and put his hands to the good earth, digging holes and putting in green cuttings, becoming "the first to plant a vineyard" (Gen. 9:20)—a moment as practical and fruitful as it was reverential.

In the same manner did conservationists approach the restoration of Cathedral Park, a modern (and very urban) vineyard. Set in the Olmos Creek watershed, this narrow strip of land lies south of the Olmos dam, between the creek's eastern bank and the limestone outcropping that forms the foundation for Torcido Drive in Alamo Heights. This fertile area has long been a landscape of plenty and promise, sustaining a rich biota. Its fecundity is what drew bands of hunter-gatherers to the area. Here they established semipermanent sites, fishing in the creek's waters, gathering wild rice from its wetlands, and hunting along its complex drainage system. Their spiritual needs were met, archaeologists surmise, as they made ceremonial use of mountain laurel seeds and other hallucinogenic material in their ritual practices.

The Spanish also found material well-being and religious sustenance along Olmos Creek and the San Antonio River it flowed into. Indeed, the presence of native peoples is what cued the European explorers to the region's beneficence. As they walked north out of the Chihuahua Desert and into the more benign landscape of what is now south-central Texas, they recorded in their journals a catalog of values. Catching their eyes first were the thick groves of elm, oak, pecan, and persimmon along the river's edge, and the fruits and nuts they produced. The explorers were also struck by the prodigious springs and abundant streamflow, which Fray Isidro de Espinosa declared in 1707, offered a shimmering prospect supplying "not only a village but a city, which could be easily founded here because of the good ground and the many conveniences." These environmental benefits—when combined with a converted workforce of indigenous labor that would, among other things, dig the fifty-mile long network of acequias to irrigate farmlands—were the bedrock for the religious missions and secular community the Spanish established downstream from present-day Cathedral Park. No wonder an eighteenth-century traveler gushed that the San Antonio River valley was "the most beautiful in New Spain."

For nearly a century the Spanish profited from this good land, as would their successors in the Mexican state that emerged in the early 1820s out of New Spain's collapse. Within twenty years, however, Mexico would be pushed out of the region with its defeat at the hands of the United States. In the mid-1840s when the Stars and Stripes rose above the Alamo, a former Spanish mission and site of the decisive 1836 clash between the Mexico and Texian rebels that had helped undercut Mexican authority, a new culture laid claim to the region's bounty.

As with its Native American and European predecessors, the United States found that its presence and power in the community was predicated on water—its flow, control, distribution, and use. For forty years the newly Americanized San Antonio continued to make use of the preexisting irrigation ditches Indian labor had constructed more than a hundred years earlier. With arrival of the railroad in 1877 and the ability to import heavy water pipes and pumps, a new water distribution system emerged in the 1880s. It did so because of George W. Brackenridge. His majestic manse, located along the San Antonio River on property contiguous with Cathedral Park, is home today to the University of the Incarnate Word. Downstream from this still sylvan setting was a pump house Brackenridge constructed, from which extended a distribution system carrying millions of gallons of water for industrial and domestic use. In time Brackenridge would acknowledge that his technological achievements, and the social benefits they produced, came at heavy cost to the riparian system that had nourished all who had lived along its banks. His subsequent anxiety about his complicity in the river's demise is a reminder of our own responsibility for its current wounded state.

San Antonio's water woes are as deep as they are many, so much so that local media calibrate every rise and fall of the aquifer's level, a handy barometer for the color of the area's lawns, the cleanliness of its automobiles, and the depth and length of the cracks in foundations, walls, and streets. Other signs of environmental distress are evident everywhere along the Olmos Creek watershed. In the early 1970s a major expressway was cut through its narrow corridor, and the steady hum of traffic and pall of carbon monoxide that hangs over it, like the fertilizers and effluent that flow down its tributaries, have turned this once fecund watercourse into an urban sink. At no time is this more evident than in the aftermath of a flood, whose roiling waters can quickly rise more than twenty-five feet behind the Olmos dam, temporarily inundating all but the basin's tallest trees. When those fetid waters finally drain, left fluttering in the canopy of oak and pecan are innumerable plastic grocery bags, tattered flags of surrender.

We need not be defeated by our behavior. The proposed reclamation of Cathedral Park, for example, offered the citizenry a profound opportunity to rehabilitate a ravaged landscape and restore a portion of a vital watershed to a semblance of its former, breathtaking beauty. The park serves as a reminder of the human obligation to repair what we have marred and of our conscious intent to nurture a greener, more sustainable city. Doing right by the land we do right by ourselves, repairing this small garden as Adam and Noah would have us do.

Springtime

By May the San Pedro Springs run dry. Their desiccated, rough limestone form, filled with crevices through which a crystalline flow rushes in wet seasons, welling up in pools that have cradled this life-giving force for millennia, suddenly looks out of place. This once vibrant environment is drained of life.

Even when mute, the springs speak loudly of their impact on human society in South Texas, reminding us of our precarious toehold in this episodically verdant land. The greenness caught the eye of the first Spanish *entrada*, which marched into the region more than three centuries ago. In 1691 Don Domingo Terán de los Rios established camp along "the banks of an arroyo, adorned by a great number of trees[—]cedars, willows, cypresses, osiers, oaks." That arboreal abundance and prolific headwaters, one missionary speculated, would sustain a city. The good don concurred. That the Payaya people he encountered at the springs "were docile and affectionate, were naturally friendly, and were decidedly agreeable towards" them only nourished his dreams of dominance. By extending Spanish military power and missionary force simultaneously from a presidio in what would become San Antonio and another on the Rio Grande, he reasoned, "different nations in between could be thereby influenced." Twenty-seven years later Gov. Martín Alarcón stood on the same ground to establish Villa de Béjar and, "fixing the royal standard with the requisite formality," made good on Terán de los Rios's imperial designs.

San Pedro Springs was also the stage where the United States enacted expansionary ambitions. In preparation for the Mexican-American War (1846–48), 1,400 soldiers drilled in San Pedro Park, which was also the construction site for the pontoon bridges they would later erect on the Rio Grande to invade Mexico. There, too, the final act of the conflict—surveying the two countries' new borders—got its start. Commissioner John Bartlett, who resided in the park while provisioning his wagon train, set off on an arduous trek from the Rio Grande Valley to San Diego to map the westward course of empire.

Within a decade this new world colossus began to fracture along a North-South fault line, the tremors of which rocked San Antonio. The city was pro-Union, a consequence of its large German population and abiding gratitude to a federal army that had protected its perimeters and produced considerable economic growth. It gave full vent to its antisecession sentiments when Texas governor Sam Houston, the standard-bearer for the state's unionists, came to town in October 1860. Accompanied to San Pedro Springs by the "city band and a large concourse of citizens in carriages and on horseback," the aged patriot received a rousing welcome from more than 2,000 citizens who crowded into "the beautiful grove," enthusiastically applauding his riveting, two-hour extemporaneous attack on "Sectionalism and Disunion."

But when Bexar County voted to secede with the state, many South Texas unionists felt embattled. Some fled for less hostile climes; those who remained bore witness to Union soldiers imprisoned in the park and listened aghast as rebels recited the Confederate constitution at the conclusion of the city's annual July 4 parade. The 1867 festivities, by contrast, featured exultant unionists waving the Stars and Stripes as they rode out to the springs, where the Emancipation Proclamation was declaimed.

The war brought some peace to the land. The city council in December 1863 passed an ordinance to "prevent the Encampment of Troops, (Ox and Mule) Trains, or other body of Men with in the San Pedro Springs Reserve and prohibiting the defiling of said Springs." Its enforcement led to the park's renovation into a late-nineteenth-century pastoral pleasure ground, scene of civic celebrations and familial rites. Yet some had greater access to these festivities than others, as the park—like the city—became increasingly segregated. When the first trolley line ran up to the park in 1878, it sparked a real estate boom that catered to the elite, whose large homes gained value by their parkland proximity. A rigid racial divide was manifest every time African Americans marched in parades ending at the springs but then were forced to picnic elsewhere; this open space was not open.

Until the late twentieth century, when the Riverwalk supplanted it, San Pedro Springs remained the communal heart, a preeminence San Antonians reenact when they gather under its massive cypress trees to commemorate Earth Day every April. And should winter storms raise the water table sufficiently to push a modest flow up through the limestone seams, we'll remember why the springs have long served as the city's rhythmic pulse.

Central Core

After slogging through Main Plaza's gluelike mud and shuddering at its squalid housing and squabbling residents, Fray Juan Agustín Morfi suspected that San Antonio de Béxar would not amount to much. It's "more a poor village than a villa capital," he wrote in 1771. San Antonians have been arguing ever since over that urban site and its communal significance.

The most recent controversy erupted in 2006, when plans surfaced for Main Plaza's reconstruction. Who decided it needed refurbishing? Why did the initial proposal close the four streets that squared up the plaza and thus scramble bus lines and other forms of transit? How come the powers-that-be did not allow the public a greater voice in the initial planning stages? And what was the precise role of the leadership at San Fernando Cathedral, whose physical form dominates the plaza's western edge, in backroom discussions with the city's elected officials, whose city council chambers are next door?

These and related issues led to rancorous public meetings. Architects argued with their peers, citizens confronted prelates and politicos, and Bexar County officials, whose courthouse occupies the plaza's southern face, were unhappy with their colleagues at city hall, as were nearby commercial interests and financial institutions. This intense clash of opinion might be the best marker of the plaza's enduring hold on the communal imagination.

Its centrality would not surprise its early-eighteenth-century Spanish designers. When they sited the plaza between the San Antonio River and the church, and then made it the pivot point for a radiating grid of streets, avenues, and alleys, they ensured it would become the town's common ground. Within its open-air expanse, religious rites began and ended, merchants hocked their wares, governing bodies assembled, courts convened, and lovers strolled.

Oh, and a horse was buried. Alive. It was entombed beneath the plaza as part of a 1749 peace treaty with the Lipan Apache, whose regular raids unsettled the isolated frontier community and its ranching hinterland. Although the truce did not hold, the burial ceremony was emblematic of the plaza's significance in San Antonio's spatial organization. This was reaffirmed across the nineteenth century when it served first as a staging ground for Santa Anna's assault on the Alamo in 1836; four years later as the locale of the bloody "Court House Fight," in which citizens and a Comanche peace party fought a pitched battle; and in 1861 as the site of federal troops' surrender to Confederate forces.

These upheavals in the town's political fortunes did not immediately alter Main Plaza's robust street life. As one visitor, Harriett Spofford, observed as late as 1877, "San Antonio is, in fact, a Spanish town today, and the only one where any considerable remnant of Spanish life exists in the United States." Yet the railroad she arrived on would quickly change the city's Iberian form and cultural feel. The east and west side depots created bustling markets and stimulated road-widening through the urban core, slicing off a portion of Main Plaza. More of its expanse disappeared in advance of the construction of an Alfred Giles–designed county courthouse in 1883.

Finally, in 1887, the city council voted to plant trees and grass in its center, turning the whole space into a park. Not everyone was happy with the beautification plans. "Leave Main Plaza open," one citizen fumed in the *San Antonio Express*. It "gives the people journeying out of the nine streets running into it a chance to turn about and shape their course to any point on the compass. We have all seen, at times, every square foot of it cut up by a wheel." His reaction anticipated poet Naomi Shihab Nye's 2007 protest by 120 years. "They should not have closed Main Avenue," she wrote in opposition to the latest redevelopment project. "A street is named Main for a reason. We needed it. For movement."

Turning the plaza into a park dealt a blow to the historic space's once vibrant, face-to-face streetscape, but even more disruptive was the arrival of streetcars and later automobiles. Rails and roads allowed the rich to flee to the city's expanding periphery, robbing downtown of much-needed energy, a process that has continued apace. Since the 1960s city planners have tried to reverse this outflow by reenergizing the plaza's pedestrian appeal through various facelifts to its infrastructure and landscaping. These efforts have not been very successful, but the most recent effort, officially unveiled in April 2008, may have better luck. It coincided with a striking increase in downtown residents living in new condos, which, when combined with the city's annual influx of tourists, may put more bodies on this hallowed ground. That influx is critical, Spanish planners understood; plazas need people.

People need plazas, too, especially San Antonio's Main Plaza, with its jumble of juxtapositions. For three centuries it has been a point of convergence and dispersal, where the citizenry of all walks of life have come together and fallen apart. Within this space, now framed by San Fernando Cathedral, the county courthouse, city government offices, and an array of businesses, treaties have been signed and pledges broken. Citizens have come here to pray, trade goods, and make money; they have made laws in a variety of languages and then transgressed them. Whether as a dirt-hardened plaza or grassy park, priests, judges, bankers, mothers, gamblers, and soldiers have claimed it as their own.

Beneath the fluttering of five flags—Spain, Mexico, the Republic of Texas, the Confederacy, and the United

States—marital vows have been exchanged and martial violence has erupted. Political dissent and religious ritual, sexual license and social constraint, mayhem and music, dance and death, like the rumble of ox-carts and trolleys, trucks, buses, and automobiles, have shaped the complex experience of Main Plaza. That's why it has been San Antonio's spatial center, spiritual core, and economic hub since the eighteenth century, and why over time its hum has proved so generative.

Going for Green

While serving as mayor of San Antonio from 2005 to 2009, Phil Hardberger rarely wasted words. But he became downright loquacious describing the 300 acres of oak savannah that, following his retirement from office, were fittingly named Hardberger Park. "For those of you who haven't seen the virgin land . . . it is truly a humbling and awe-inspiring experience," he confessed in his 2007 state of the city address. "It may even move you to hug a tree, though you won't be able to get your arms around many of them." Their girth and age provoked a nostalgic reverie. "To walk among those trees—many older than the heroes of the Alamo—is to know our history," he enthused. "It is a breathtaking expanse of urban wilderness."

Although the north-side property was neither pristine nor wilderness—it had been a working farm and dairy for generations—the mayor's more important insight was spot-on. The park offers the citizenry an unparalleled chance, even if only momentarily, to step outside the mad rush to pave over every square inch, which has flattened San Antonio's contours and homogenized its vistas, filling each with more shopping centers, freeways, and subdivisions. Against this jarring backdrop the park would become an oasis, Harberger predicted, "and it is now ours to keep."

How the community would keep its quiet, sylvan beauty, set in a booming metropolis, was of vital concern. The city's park inventory in 2007 provided only 14.5 acres per 1,000 residents, 2 acres per capita less than the national average. The gap used to be wider; forty years ago San Antonio had less than half the national average. And while the community has made up a lot of ground, in future decades growth in parkland acreage might be more difficult to come by. According to the 2000 Census, San Antonio experienced its first increase in density since 1950, a pattern that will continue if the regional population hits the 2.5 million predicted for 2040, adding greater pressure to the area's already stressed parks.

This demographic projection is why it mattered how carefully and consciously Hardberger Park (and subsequent open space) would be designed. Believing that this bucolic parcel could become a signature landscape, with an impact akin to the city's nineteenth-century green gem, Brackenridge Park, San Antonio launched an international competition to entice renowned landscape architects to bring their perspectives to bear on the park's varied landforms and complex of ecozones. I was a member of the jury evaluating their proposals and was blown away by the diversity in design and detail, from the clever treatments of terrain and topography to the artful use of native flora and local culture. These outsiders understood San Antonio in ways locals often miss about themselves and the places we turn into home.

The park design was ultimately entrusted to Stephen Stimson Associates and D.I.R.T. Studio. Their creative thrust was announced with their initial vision statement. It would be "a cultivated wild at the edge of San Antonio." Imagining a landscape integrating natural forces and human stewardship, the firm proposed to cultivate indigenous grasses, reinvigorate oak woodlands, and restore riparian flows. They also planned to pay homage to the land's original owners, Max and Minnie Voelcker, and their dairy operations by rehabbing farm structures and constructing an education center to teach visitors how to live more deliberately on the land.

They proposed testing these compelling aspirations through an open and accessible design process. In November 2007 the city and neighborhood groups sponsored a series of public meetings at which the designers presented their conceptions of the park, listened to the community's reactions, and returned to their drawing boards. This first stage, like those that followed, allowed the citizenry to practice the kind of participatory democracy so crucial to the creation of a more habitable city.

Hardberger Park by itself will not reverse San Antonio's shabby legacy of too few playgrounds, soccer fields, and hike or bike trails, and its diminishing forest canopy, farm lots, and grasslands. It did not contain enough in

1950 for a city of 408,000, and it still does not with three times as many people in the early twenty-first century. But the latest addition to the city's inventory nonetheless marked a watershed moment in local park history, sparking a more proactive pursuit of recreational landscapes in a community that only a century ago touted itself as the "City of Parks."

The nickname implied a resolute commitment to parklands development that San Antonians knew was not true; they inherited most of the available open space. Innovative eighteenth-century Spanish urban planners had designed a street grid centered on Main and Military Plazas and set aside San Pedro Park. In 1899 George Brackenridge built on this open-space legacy when he donated the first 199 acres of what is now the 340-acre Brackenridge Park. A third wave of land donations across the early twentieth century were spin-offs from or come-ons for real estate schemes, including Woodlawn Lake, Collins Park, and a clutter of smaller odd lots that developers have sold to or dumped on the city.

That pattern changed substantially in the 1920s when park development became a political force, courtesy of a series of city bonds that Parks Commissioner Ray Lambert aggressively packaged and promoted. His activism led to the purchase of nearly 1,100 acres, a figure unmatched for another four decades. Most were located on the east side, a nod to African American voters whose overwhelming support for the bonds contrasted sharply with no-tax sentiment on the white north side. Not until the 1960s did San Antonio invest significantly in parks to enhance quality of life, yet even the 1,563 acres it then purchased—mostly to create McAllister Park near Interstate 410, which was the outer edge of development—seemed lost as the city exploded in size.

Since then the city has been trying to catch up with the swift movement of car-driven sprawl. But despite the passage of propositions in 2000, 2005, and 2010 to use sales tax receipts to purchase land over the Edwards Aquifer's sensitive recharge zones, and notwithstanding the designated parklands built into the urbanist-inspired development known as City South, San Antonio has been slow to dedicate itself to a well-funded campaign to build and endow a dense, equitably distributed network of parks so that its many residents can make full use of the peace, beauty, and breathing room it would offer.

Should Hardberger Park generate a more sustained commitment to a greener and more playful future, San Antonio might distinguish itself, in its namesake's words, as "one of the great livable cities of North America."

Ebb and Flow

Ronnie Pucek had San Antonio over a barrel. Admittedly it was a barrel of catfish, but that only made the situation more comic and contentious. In 1991 he and his silent partners purchased land in south Bexar County, drilled a monster well, and, making ready use of Texas's arcane right-of-capture law, began pumping more than 40,000 gallons a minute to flush through the Living Waters Catfish Farm. Although there was no legal constraint on the amount of water the aquaculture concern could use, its impact on the Edwards Aquifer—and thus on San Antonio's potable supplies—set off alarm bells. The city and the San Antonio Water System scrambled to find grounds to halt Pucek's operations, noting that no single entity should have the right to absorb as much flow as could support one-quarter of San Antonio's population. Over the next ten years they spent buckets of cash on legal fees, and in the end they bought out Pucek and his backers for approximately $30 million.

This was a very expensive lesson for the city, the first of many during the 1990s, which was arguably a pivotal decade for water politics and policies in San Antonio's modern history. Alas, in recent years the city has ignored some of what it learned—one reason it was recently identified as the country's fourth-largest city that is running out of water.

It's not as if San Antonians have been unaware of the significance of water, in all its forms, to life in south-central Texas. In 1891 banker George W. Brackenridge funded an exploratory well on Market Street, and its prodigious gush launched a cheap-water era that lasted a century. There was the flood of 1921 that tore the town apart. Deserving mention, too, is the searing drought of the 1950s. And there have been some exquisite brawls over the Edwards Aquifer's vital recharge zones. In the 1970s developers and activists tangled over housing subdivisions and shopping centers, struggles that have been reprised ever since.

Each of these stories, however, found full and interlocking expression in the last decade of the twentieth century. As a result San Antonio had to develop—however painfully and slowly—a new consciousness about water, a new way of inhabiting its home turf.

Start with floods. What the native peoples knew, the Spanish came to recognize and the National Weather Service now confirms: San Antonio lies in Texas's Flash Flood Alley. Powerful storms routinely rock the region, dropping immense amounts of rain, like the nineteen inches that slammed down in 1998. Once that amount would have scoured the downtown core, but this time its torrent was captured by the San Pedro Creek and San Antonio River tunnels, completed in 1991 and 1997 respectively—$150 million that was well spent.

When deluge gives way to drought, as happens when the weather pattern shifts from El Niño to La Niña, so do the environmental issues that wrack the community. The bellwether here is the still flowing Comal Springs, one of the few remaining in the region. Most of the others dried up due to excessive groundwater pumping. But because those in New Braunfels hung on into the 1990s, they became embroiled in a landmark lawsuit. In 1991 the Lone Star Chapter of the Sierra Club used provisions of the Endangered Species Act to sue the U.S. Fish and Wildlife Service for failing to protect the seven endangered species that live within the Edwards Aquifer's fluid embrace. The San Antonio political establishment went ballistic, knowing that if the lawsuit succeeded, the Alamo City would be compelled to change its hitherto wasteful water regime. In a profound failure of judgment, it refused to plan for the inevitable, and when the court sided with the Sierra Club, it had wasted precious dollars and time.

The same happened with its ill-advised efforts to build the Applewhite Reservoir in southern Bexar County. After a decade-long permitting process, voters rejected the project in raucous special elections in 1991 and 1994. These back-to-back defeats forced the city to consider a longer-term water strategy that would meet federal regulatory requirements and provide a sustainable supply of water.

The process became more compelling and complex with the creation of the Edwards Aquifer Authority, which the state established in response to the Sierra Club victory in federal court. Finally, a regional agency would manage the aquifer's resources, a shift of unparalleled significance. An example: its emerging rules and regulations liberated the San Antonio Water System (SAWS) to pursue a robust conservation agenda. It raised water rates in the late 1990s and then used the capital to educate consumers about low-flow technologies and offer financial incentives to homeowners to make the switch. It also spent millions on a water-recycling plant whose reclaimed flow irrigates parks, golf courses, and campuses and began planning a state-of-the-art aquifer storage system that would ultimately cost $110 million. Each of these bold ventures saved lots of water and significantly reduced pressure on the aquifer and the species that inhabit it.

Another element in SAWS's conservation strategy has been more problematic. Consider its commitment to water ranching—that is, buying up water rights from ranchers, farmers, and others to supplement the city's supplies. This tactic is what netted Pucek's catfish farm, a good outcome. Much less beneficial were subsequent efforts to sop up rights throughout the Hill Country. SAWS has not built pipelines to these sources but instead pumped newly purchased "paper water" out of its wellheads in the city. This underground transfer of its liquid assets has intensified pressure on the Comal spring flow—

the same situation that produced the momentous Sierra Club lawsuit some twenty years ago.

Let me put it this way. In the 1990s SAWS and the city of San Antonio shucked their longstanding reputation as one of the worst water hogs in the West and became smart, aggressive, and forward thinking. Since then this innovative energy has flagged. Worse, with the exception of the land purchases over the aquifer recharge zones, SAWS and the city have backslid to such an extent that the time seems ripe for another legal challenge.

Should Sierra Club II occur the city will lose again. The federal government will force it to stop pumping paper water and begin piping it in from the Hill Country. The good citizens will be compelled once more to pick up the tab because ten years ago their political leaders and policy wonks hit the snooze button.

Somewhere Ronnie Pucek must be laughing.

Back to Nature

Coyotes were returning to Olmos Basin, millions of American snout butterflies were on the wing, and an opportunistic palm tree had planted itself next to a telephone pole north of the McCullough-Hildebrand intersection. Each is a striking reflection of the integration of the natural and human landscapes, a weave so tight that it often escapes our notice or comprehension. But if we want to make our presence on this good land more sustainable, we must pay closer attention to the dynamic terrain that encompasses all life in south-central Texas.

This claim about the human place in nature carries an assumption that must be acknowledged. No one understood better than Henry David Thoreau the need for self-consciousness, and he urged readers of *Walden* to "Explore Thyself" as the essential first step to enlightened action. That is what can happen when you walk into the woods to think.

Or when you eat an apple. Just ask Adam and Eve. They were booted out of Paradise for knowing themselves (and each other), and we haven't stopped talking about the price they paid for that delicious bite into self-awareness, making their story a source of deep-seated tension in Western culture. Because they fell from grace, humanity was expelled from Nature, an Eden we cannot regain; separated from the physical world, we defile what we touch. Because our every act is a sacrilege, we are aliens on earth, our home ground.

Our alienation reached such unacceptable levels in Genesis that divine punishment ensued, first in a devastating flood and later in the destruction of the Tower of Babel and the human self-regard it was thought to reflect. The tautology these Biblical stories have spawned makes it clear there is a line humans cannot cross.

Yet paradoxically we must cross that divide if we are to escape the devastating consequences visited upon us by a Biblical conceit pitting us against nature. If the human can never enter the wild and the wild can never be human, there is no reason to feel responsible for any species except our own or to embrace any landscape other than those we have constructed. By this self-referential logic, no coyote, butterfly, or palm has standing.

Naturalist John Muir knew better. While trekking a thousand miles from Wisconsin to the Gulf of Mexico in 1867, he developed this insight: nature and its plentitude operated not for human gratification but for their own ends. With this appreciation he put humans in their place, less exalted and more alert. The demand for an enhanced sensitivity shaped Progressive Era thinkers, too. Alice Hamilton and Jane Addams recognized that the industrial city and its often impoverished inhabitants required a new kind of environmental analysis that took seriously the urban challenges that came from extended contact with tainted water, befouled air, and squalid housing.

Whether citified or rural, conservationists must get their hands dirty. So argued ecologist Aldo Leopold in his 1949 *Sand County Almanac*. The best definition of their ethic "is not written with a pen, but with an axe. It is a matter of what a man thinks about while chopping, or while deciding what to chop." These internal contemplations have external ramifications, he wrote. "A conservationist is one who is humbly aware that with each stroke he is writing his signature on the face of his land." Only when we admit that our actions have consequences and that we cannot *not* act will we be able to embrace an environmentalism that seeks "a state of harmony between men and land."

Pushing for exactly this result was Gifford Pinchot, first chief of the U.S. Forest Service. In the 1930s, as governor of Pennsylvania, a heavily industrialized state reeling from the Great Depression, he promoted a program ameliorating the twin scourges of those hard times—human tragedy and environmental catastrophe. Confronted with escalating unemployment as the mines and mills shut down and faced with thousands of acres of badly cutover timberlands that logging companies had abandoned, Pinchot put people to work replanting the battered terrain with the goal of "making Penn's woods,

woods again." Social justice and landscape restoration went hand in hand.

Not all public policy was as perceptive. Take DDT. In her 1962 book *Silent Spring*, Rachel Carson tracked its poisonous path through air and water and with terrifying ease revealed how it altered genetic structures of flora and fauna, including the humans who invented the deadly carcinogen. By our every inhalation and swallow, we affirm our ecological integration, our earthly bonds.

Communities are also bound up in their environs. Certainly San Antonio could not exist without the prodigious Edwards Aquifer and the recharge features that created its cavernous extent, the springs from which have bubbled billions of gallons of water, and the rivers that bear away its life-giving force. The hunter-gatherers who roamed the Balcones escarpment for more than 10,000 years recognized this, as did the Spanish who arrived in the late seventeenth century. That reliance has remained even as the Alamo City evolved from a dusty, compact frontier outpost to a sprawling Sunbelt megacity.

The challenge is to bind together this robust natural system and the human ecology that is utterly dependent on it. From landscape preservation to aquifer protection and wetlands restoration, from green building to alternative energies and smart growth, we must better meld the community with its commons. The task for some will be to punch holes in the impervious cover, making a more porous city. For others it will be to extend the city's park system, enhancing recreation and aesthetics. For still others the goal will be to recenter San Antonio, luring people back to the in-town neighborhoods so many fled to chase the suburban dream and decreasing their intense reliance on the automobile that made sprawl possible. No less critical is the conscious pursuit of social justice. Demanding a living wage for all workers, boosting their and their children's educational prospects, and constructing affordable homes, like the crying need to develop a more diverse economy, are among the hurdles we must clear to ensure public well-being and ecological health.

Let's face it. These goals will not be easy to achieve, and even if we are lucky, they will take generations to be realized. Like the graceful river that wends its way through this 300-year-old city, San Antonians' efforts will have no real beginning or end. They must be as enduring as the river's relentless reach to the sea.

TWO Rough Waters

No Relief

He woke with a start. "I felt something cold, looked down and there I was with water in my lap," reported a man who had drifted off as the storm raged. What had been outside was now inside, and when he pushed through the door to escape the dark waters surging into his house, he was stunned. "God, it was like one giant swimming pool as far as the eye could see. There were people I knew—women, children, screaming, praying . . . A woman who lives down the block floated past me with her two children beside her." Others had loved ones ripped from their arms. A father who struggled to hold onto his five children as Lake Pontchartrain roared into the city mourned, "I couldn't do it. I had to let two of them go."

With its levees breached and infrastructure torn apart, with power gone and effluent coursing along once dry roads, sweeping up corpses, automobiles, and the odd runaway boat, New Orleans was caught in a tangle of flotsam and jetsam. Fetid and foul, looters donned stolen scuba gear to evade police surveillance, helping turn the tourist mecca into an urban nightmare.

This description of New Orleans sounds much like

what happened in the aftermath of Hurricane Katrina in 2005, when New Orleans and the coastal communities blacked out and went underwater. As causeways and communications collapsed, as the death toll climbed and the social fabric frayed like the Mississippi Delta slipping beneath the rising tide, the gruesome images of devastation provoked a deep sense of disbelief. But the account was in fact about the catastrophic collapse of the city following the September 1965 landfall of Hurricane Betsy. It slammed into southern Louisiana, splintering coastal communities, submerging tankers and barges, swamping oil refineries, busting pipelines, and flushing sewers, and then made a mockery of the levee infrastructure. Betsy, like Katrina, left the Crescent City to flounder in its filth.

That these two storms were so similar is not by happenstance. New Orleans is deeply entangled in the enveloping natural world, as reflected in its sobriquet, Crescent City—a name derived from the river's bend that makes New Orleans possible. Its residents have expended immense amounts of human energy, social capital, and money to make (and keep) dry what long had been wet. Yet in the process of seeking an ever-greater distance between itself and the river, lake, wetlands, and the Gulf, the city's actions have imperiled public safety. The canals, channels, and levees designed to lift New Orleans above its swampy base have intensified its subsidence; automobiles have driven the construction of new, outlying, and below-sea-level subdivisions, accelerating environmental degradation and elevating the risk to life and property. Every new development brought the need for increased protection, an endless cycle.

Look what happened after Betsy. Help arrived quickly. President Lyndon Johnson landed within twenty-four hours, touring the devastated community "to see with [his] own eyes what the unhappy alliance of wind and water has done to this land and people." His generosity of spirit was backed up by federal largesse; within a month Congress had appropriated $250 million for Louisiana, 20 percent of which was targeted for New Orleans's complex levee system. It should be noted that there was a little self-help in this massive outlay. The storm "had landed hard where an influx of Texas investors, Lady Bird Johnson among them, planned to levee off 32,000 acres for 250,000 people in a new suburb and industrial park," writes historian Todd Shallat. And so in this case, and all others, the city was rebuilt, its economy fueled by petrodollars, convention cash, and tourist leavings, a buoyancy that depended on thicker levees and higher embankments, concrete channels, and state-of-the-art pumps. Yet this set of technological responses that "brings prosperity and security to humans is literally costing them the earth beneath their feet," one New Orleans resident confided to Shallat.

The title of Craig Colten's haunting book, *An Unnatural Metropolis: Wresting New Orleans from Nature*, reflects the reason for this. Colten makes clear that for nearly 300 years an immense amount of human labor (slave and free), along with equal parts of human ingenuity and idiocy, have been manifest in the incessant effort to plan, construct and reconstruct, and configure and reconfigure the wetlands and bayous the city was first platted in.

Naturally the French are to blame. In 1722 they identified the most elevated terrain they could find along the lower Mississippi, once a Quinipissa village, as the most suitable location for a new entrepôt. Introducing cattle, rice, and slavery to the landscape, French farmers altered its sustainability. Rice production demanded a pliable river, precisely what the Big Muddy was not, and stable riverbanks, which grazing cattle weakened. To strengthen the levees that were required for the colonial experiment to endure, slaves were put to work with shovels and pickaxes, building flood-control structures even as they increased the acreage under cultivation, further threatening the levees' ability to protect the expansion-minded settlers.

That logic, frequently self-defeating, has resurfaced time and again. What would happen, for example, when levees were constructed in one area but not another? Floodwaters were pushed into unprotected zones and then swirled around and behind the leveed terrain. To prevent a reoccurrence, more barriers were erected, and by the mid-1730s a forty-four-square-block area was completely circumscribed, giving its inhabitants an eagerly embraced sense of security. Their newfound faith in the engineered solution collapsed—if only temporarily—under the weight of a 1735 flood that burst over the walls, inundating the community. The rush to build higher levees, which triggered demands to extend their reach up and down the river along both banks, accelerated regional water woes. The walls rose from four feet to six, enclosing high waters in a smaller, narrower space, and flood levels rose proportionally, as Spanish New Orleans discovered in 1785 when a massive surge crashed over and through the earthworks.

The region's levees-only policy remained unchanged when the United States occupied Louisiana in the early nineteenth century, although in 1846 the state engineer, P. O. Hebert, raised serious doubts about its ramifications. Believing that higher levees increased "the danger to the city of New Orleans and to all the lower country," he proposed instead: "We should . . . endeavor to reduce this level, already too high and dangerous, by opening all the outlets of the river. We are every year confining this immense river closer and closer to its own bed—forgetting that it is fed by over 1,500 streams—regardless of a danger becoming more and more impending." Few listened, even after the disastrous 1849 flood, in which more than six feet of water covered 200 square blocks.

Nor did any in the drenched community pay much attention to the social inequalities revealed by the repeated flooding. As Colten notes, there is a direct relation between wealth and water. The poor got wet, and the rich remained dry (or drier). In 1849, for example, more than 12,000 inhabitants of the city's tenements became refugees or, as one contemporary account put it, were forced to "live an aquatic life of much privation and suffering," an amphibian existence that was their lot again after the 1890 flood.

The disadvantaged also suffered disproportionately from disease, carried by mosquitoes that thrived in the swamps and stagnant waters, above which New Orleans rose barely at all. (As Colten argues, the levees themselves are the only visual relief in the city's otherwise flat landscape.) Yellow fever, for one, was especially prevalent in impoverished backwater neighborhoods, not incidentally those areas with the least steep grade and thus the most clogged drainage ditches, a topography that created a network of "open-air septic troughs." By the late 1930s a WPA-funded sewage system went online, but not everywhere. The lowest-lying, poorest, and African American wards remained unsewered. This deliberate policy of exclusion reflected the city's long-established pattern "of turning low-value land associated with environmental problems over to minority populations," thereby intensifying the deprivations visited daily upon those least able to escape them.

Not much has changed in the succeeding seventy-five years. Still sited on sodden ground; still segregated, and tightly so; still vulnerable to the punishing blows of Gulf storms and river floods, yet with "situational advantages" that are undeniable—in the nineteenth century

New Orleans was the gateway to the North American interior, and today one-third of the country's oil flows through southern Louisiana—the city hangs on for dear life. How much longer it can do so is anyone's guess. But a source of its potential demise seemed clear enough to Colten when he wrote the epilogue to *An Unnatural Metropolis*, published eight months before Katrina crashed into "the city care forgot."

Colten noted that its increasing physical instability, as measured by its steady subsidence, when linked to its proximity to the "massive and shallow Lake Pontchartrain," places the community in dire straits.

> Should a Class 5 hurricane blow water over the lakefront levees, the city could find itself under water for months. Evacuation would face serious bottlenecks due to a limited number of escape routes across the waterlogged terrain—and some of those raised highways could be over-topped by storm-driven waves. Recent popular accounts paint a dire picture and suggest that federal authorities might not be willing to make the investment necessary to save a city that cannot protect itself. Global warming and sea level rise make this grim forecast all too likely.

In Katrina's immediate aftershock, Colten's insights proved eerily prescient. While our Nero of a president fiddled, House Majority Leader Dennis Hastert (R-IL) wondered whether repairing New Orleans made sense. It did not, at least to one deeply frustrated resident trapped in the hellish Superdome. "We've been sleeping on the (expletive) ground like rats," fumed Marc Levy to an AP reporter. "I say burn this whole (expletive) city down."

Water, not fire, may do his bidding. Of the many aerial shots of New Orleans that became the staple of television news reports in the weeks following Katrina, among the most potent were those filmed by dawn's early light. The city glistened, and a silvery sheen rippled along its once fabled streets and lurid back alleys, resembling nothing so much as the wetlands they were built upon. Nature had reclaimed its own.

Storm Warning

Disaster relief. The words have a reassuring sound and offer a sense of hope—everything will be all right. But as South Texans discovered in the 2008 wake of Hurricane Dolly, just as Gulf Coast residents learned in the traumatic aftershocks of the 2005 monster storms Katrina and Rita, it takes a long time for life to return to a semblance of its former self.

Recovery is especially hard and made more complicated when major hurricanes repeatedly sweep through the Gulf of Mexico, hammering the coastline. Cleaning up after another set of screaming winds and pounding squalls is only part of the effort required to make things right. The psychological stress of being on the receiving end of another big blow is an in-your-face reminder of how vulnerable coastal life can be.

No amount of federal, state, or local funding—assuming it is forthcoming—can recapture all that has been lost, physically and mentally. Dolly's victims know this well. The $24 million in federal aid that flowed into the Rio Grande Valley in the hurricane's aftermath was a drop in the bucket. Although it shored up battered infrastructure, helped South Padre Island hotel and condo owners nail down new roofs, and provided essential short-term unemployment funding, its total fell considerably short of the $1 billion needed to fix the region.

In March 2010 Jesus Flores, who lives in the San Carlos colonia, went public with his grievances. Eighteen months after Dolly ripped apart the ditches that drained his street, flooded his home, and compromised his health, little had changed. The infrastructure remained in disarray—"They're supposed to come and clean it. They say they are going to come. For two years they never came"—and this neglect for public works was mirrored in the broader failure of local, state, and federal authorities to aid in the recovery of the lives and livelihoods that Dolly disrupted.

Even if governmental intervention had been more robust, its impact might still have proved only cosmetic. The infusion of federal dollars, for example, suggested that the only issue South Texans needed to focus on was how to pay for the physical rebuilding of damaged places. It deflected policymakers and individual citizens from considering more carefully the larger environmental ramifications and geopolitical implications of Hurricane Dolly's surge across the borderlands.

Nature has no respect for political geography. It does not pay attention to national boundaries or cross-border differences in economy, language, and culture. It could care less about the social, ethnic, or racial categories that mean so much to human beings. It is indifferent to the walls we erect. We are likewise frequently unmindful of nature's power and import, though that's not a particularly wise strategy. Whenever we get too comfortable with our putative capacity to dominate the world or fool ourselves into thinking we are immune to nature's vicissitudes, this good earth reminds us how dangerous our hubris or inattention can be.

One of those blunt reminders came with Hurricane Dolly's swirling power. As it smacked into South Padre Island on July 23 and slowly spun up and along the Rio Grande Valley, the National Weather Service radar tracked its fury; its multihued imagery captured the storm's vast size, its jagged movements, and the intensity of its winds and rain. Although it only reached Category 2 status and did not pack the wallop of Katrina or Rita, its gale-force winds, high tides, and tornadic whirl left a devastating trail of destruction.

Dolly's ravaging energy kept me glued to my computer. In part that's because I am a storm junkie. I have experienced a number of hurricanes in New England, from Carol (1954) to Bob (1991), and I lived in San Antonio for the deluges spun off by Gilbert (1988) and Bret (1999). So I had a feel for Dolly's churning, erosive surge crashing on the barrier islands, the slick and dangerous streets, the overturned cars and debris-filled sky. Technology helped me see what I could not witness firsthand, to remember what it felt like when the lights went out, day became night, and the air became electric.

I was also transfixed by the animated, glowing Doppler radar. It conveyed a critical lesson about the collision of international politics, Mother Nature, and human geography. Dolly's eyewall may have made landfall on the lower coast of Texas, but its radiating force pummeled Americans and Mexicans alike. Its 100-mile-per-hour gusts ripped up trees, blew off roofs, punched holes in buildings, sank boats, and dropped power lines on both sides of the border. As rain rushed into the Rio Grande's many tributaries, the rising waters swept into neighborhoods and barrios alike. With funnel clouds skipping across the borderlands, life was disrupted in Brownsville as in Matamoras, in Reynosa as in McAllen. Whether seen from an orbiting satellite or witnessed on treacherous terrain, Los Dos Laredos became one.

We should remember that shared moment, even though buildings were quickly pumped dry, shattered windows were replaced, and the thick tangle of trunks and branches were cleared away. We should recall the vivid sense of sameness that Dolly generated, even though construction crews immediately resumed building the "security fence" separating those on the Rio Grande's northern banks from those on its southern. However much their labor implied a reclaiming of order out of chaos, the on-the-ground consequences of this activity appeared parochial, insignificant, and narrow-minded when set against nature's stunning capacity to obliterate differences. We have acted as if this fence, a bit of territorial marking, could restructure the world we co-inhabit with those who happen to live on the other side. We assumed that this human-made boundary, this line on a map, could protect us from forces beyond our control.

Ike's Wake

Just before President George Bush boarded a helicopter in September 2008 to survey the devastation Hurricane Ike wrought on the Texas Gulf, he gave an upbeat appraisal of the flattened landscape. "It's a tough situation on the coast," he asserted. "I've been president long enough to have seen tough situations, and I know the resilience of the people to deal with tough situations. I know with proper help from the federal government and the state government, there will be a better tomorrow."

He sounded just as tough after Katrina savaged New Orleans in 2005, promising that it too would rise again. It hasn't. Which is why residents of Galveston and the obliterated Bolivar Peninsula were appropriately skeptical of this latest declaration of presidential blue-sky optimism. Ever since, there have been plenty of dark days.

How dark depended less on the president's words than on how critically Galveston assessed its site and situation. In deciding to rebuild in the same footprint, in assuming that technological innovations—bigger bulwarks and higher stilts lifting homes above presumed storm-surge lines—would fix its precariousness, by concluding that tons of new sand were necessary to re-create the beaches Ike ground away, and by flogging the refrain on the city's tourist website "The Island has seen its share of calamities, yet the worst natural disaster in U.S. history could not erase the tranquility of a Galveston sunset," it was asserting that the environment must bend to human will. In so doing, it was setting itself up for yet another fall.

There is a reason, after all, even a midsized Ike was able to tear Galveston apart, leaving behind a splintered community and a distraught people. Built on a barrier island with little natural elevation or solidity, the city of more than 57,000 residents, which attracts millions of tourists a year, never had a chance. Its odds of surviving a direct hit were so small because of how and where the city was built. They shrank further in part because of its inhabitants' stunning misjudgment. Shell-shocked survivors confessed to reporters that they had faith in the city's seawall; they were certain they were prepared. This post-Katrina confidence seems misplaced at best. At worst, given Galveston's long experience with deadly hurricanes, it seems suicidal.

One of the few things Ike did not sweep away was an arresting sculpture that had been bolted to the historic seawall. The simple bronze statue, erected in 2000, consisted of three figures locked in close embrace—a father, mother, and child sunk into rising water. The father held fast to his family with his left arm; his right was thrust into the air, fingers outstretched in a signal of distress.

The sculpture memorialized the human plight when an unnamed Category 4 hurricane blew the city to pieces in September 1900, killing more than 6,000. Considered the worse disaster in American history, that storm destroyed Galveston's soul.

Ike, which clobbered the island almost exactly 108 years later, did not smash the statue, but in the broken city we can read a larger warning. We need to rethink our concrete presence on the coastal Gulf in an age of supercharged hurricanes, much as the Japanese people must in the wake of the 2011 killer tsunami that wiped out thirty-five-foot seawalls before sweeping away everything else.

Hurricane Ike exacted a huge toll because of the collision of two forces, one climatic and the other demographic. The slow but steady increase in the earth's temperature, which has caused a rise in sea level, has had a powerful impact on the development of hurricanes. These storms, born in the warm waters off Africa, spin across the South Atlantic and refuel in the bathtub-warm waters of the Caribbean Sea and the Gulf of Mexico, swelling in size and ferocity. Their extent—Ike reached epic proportions, with a cyclonic swirl 600 miles wide—has allowed them to start pounding islands and the mainland long before their full force reaches these compromised coastlines. As their surges eclipse barrier islands and beaches, they grind up these natural defenses and undercut any concrete structures designed to repulse smaller storms and lower waters. The effects of climate change are manifest in every eroded beachhead, flattened mangrove swamp, and drowned wetland.

Human dangers soar, compounded by the rapid increase in the region's population. Run your eye east along a map of the Gulf Coast: Brownsville, Corpus Christi, and Galveston; New Orleans and the Florida Panhandle; Saint Petersburg, Tampa, and Fort Myers. These cities form an arc of explosive growth along the Third Coast. Young and old have flocked to these warm and humid landscapes, boosting successive waves of migration to the Sunbelt that commenced fifty years ago. The increased value of oceanfront real estate generates the construction of untold numbers of apartment buildings and condominiums that hug the shore. More people are crowding into these low-lying coastal enclaves when their glittering sands are most imperiled.

That is why when Hurricane Dolly slammed into the southern Texas coast in July 2008, its damages topped $1.2 billion; why Gustav's punishing run through the Caribbean later that summer, which ended only after it ploughed into Louisiana, wracked up more than $20 billion in losses; and why early estimates for Ike's crushing journey were close to $30 billion. The loss of human life is incalculable, but we know for certain that as more of us live in harm's way, more of us—and the communities we inhabit—will come to harm.

Mounting a sustainable defense of these swelling population centers amid global warming becomes a complex, perhaps impossible task. If a Category 2 storm such as Gustav, which veered well west of New Orleans, still managed to overtop the city's levee structures, how high must we build to resist a direct blow of similar or greater magnitude? If Ike, another Category 2 storm, could collapse Louisianan levees far to the east of its Galveston landfall and simultaneously rush over the Texas city's seventeen-foot seawall, what would it take to repel a bulldozing Category 3 storm (or worse)? How much concrete are we willing to pour? How much money are we willing to spend, and who decides? Who benefits?

These are just some of the crucial policy questions we must answer if we want to construct a genuine, life-sustaining memorial to the grief-stricken residents of the Gulf Coast. The last thing they needed, as they tried to pick up the pieces of their shattered lives, was mere tough talk.

THREE Borderline Anxieties

Fiery Deaths

Areli Peralta. Ruben Santos. Lourdes Tadeo. Jose Israel. Maria Beltran. Juan Bautista. Each was a victim of the October 2007 Harris fire that erupted along the Mexico-California border. It burned more than 90,000 acres over twenty days, killed at least eight people and injured hundreds, and led to massive evacuations throughout San Diego County. By any measure the Harris fire was as ferocious as it was deadly.

Seen from satellite photographs, the fire's smoke plumes obscured northern Baja and Southern California and streamed far across the Pacific Ocean. Its flames paid little attention to political boundaries, leaping across fences the United States has erected to stop migrants heading north and at one point forcing border agents to close the Tecate port of entry. The dividing line the federal government has tried to control through high-tech surveillance equipment, security walls, and night-flying helicopters instead fell under nature's fiery dominion.

Some took advantage of the disruption, cutting the chain at the Tecate entryway and hustling into the United States. Among them were Peralta, Santos, and Beltran. They and others found themselves in an arid brush country engulfed by wind-driven walls of fire. Survivors told reporters they knew they could not outrun the blaze. Some lay down in streams, others crouched behind boulders, and still others tried to get above the fire, scrambling up the rugged terrain in search of safety.

Those who could, however scorched, bruised, or gagging, turned back. U.S. border officials arrested more than 200. Dozens more were rescued after the fire died down. These men and women had paid a steep price in pursuit of their dreams.

No one appreciated the migrants' plight more than a group of Tijuanan *bomberos*, firefighters who joined the battle against the Harris fire. They spotted one of the bodies and reflected on the great disparities that have led so many to take such risks to reach El Norte. "It's the consequence of the United States being a First World country and that Mexico is not. It's sad," Rodrigo Santana told the *Los Angeles Times*. Juan Carlos Mendoza said, "We worry about them and it pains us because they're Mexicans and they're one of us."

The sixteen who survived despite their massive burns spent painful months in rehabilitation at the University of California–San Diego Regional Burn Unit, receiving treatment costing an estimated $1 million. That these survivors, undocumented immigrants, received any medical care, let alone that it proved so costly, angered some San Diegans, igniting a political firestorm. The hospital was barraged with calls from people enraged that taxpayer dollars were being spent on the patients. A year later people were no more forgiving. The *Los Angeles Times* retrospective series on the burn victims was, one correspondent snorted, a collection of "illegal sob stories."

Amid the death, destruction, and controversy lay a hopeful sign of transnational amity. The Tijuanan *bomberos* were thought to be the first such contingent to cross into the United States to fight alongside their American counterparts. Their efforts were invaluable, asserted August Ghio, fire chief of San Miguel, California, proving that there are "no borders when something like a fire or a catastrophic natural event happens."

The real tension, however, is not political but environmental and economic. As I write this, high winds have started to rip across southern California and northern Baja, sparking small fires. Should this natural cycle flare up as it did in 2007 and 2008, should it intersect once more with migrants hoping to slip across the border seeking work and a better life, we may bear witness to another human tragedy along the borderlands fire zone.

Lockup

When it comes time for historians to determine George Bush's presidential legacy, they will start by addressing his vainglorious prosecution of the Iraq war, the torture he sanctioned at Abu Ghraib and Guantanamo prisons, and his domestic power grab that granted the chief executive unconstitutional authorities. As shocking as these extensions of his clout were, they were particularly stunning for the way they subverted his electoral claim to be a devout advocate of small government. As president he created the most intrusive big government Americans have experienced since Richard Nixon. Both presidents placed themselves above the law. As Nixon famously concluded and Bush's behavior confirms, "When the president does it, it means it is not illegal."

Scholars will have a field day probing these unsettling parallels and assessing Bush's unique demerits. When they do, I hope they will also dig into something a bit more down to earth, like the significance of 1.7 million cubic yards of dirt—or the 35,000 truckloads that were required to move it. This, too, is a big story that says a great deal about President Bush's abuse of political and physical power.

As part of his plan to enhance America's border security, in 2005 the president claimed he had the authority to set aside individual state and federal environmental regulations. He then proceeded to do so, ordering Homeland Security to build tall, triple-thick fences through public lands and private property, bisecting wildlife refuges, walling off communities, and cutting up ranches from Texas to California. Perhaps the most egregious example of this presidential heavy-handedness occurred along a three-mile stretch of the border between San Diego and Tijuana. The administration's ambitious goal was nothing short of the reconstruction of the entire terrain.

Ignoring environmental protection lawsuits and the California Coastal Commission's objections, the federal government spent more than $60 million in taxpayer money to level hills and mesas, fill in Smuggler's Gulch and Goat Canyon with the soil, and construct a 150-foot-high berm to seal off the smoothed-over landscape. Topping the imposing earthen structure are stanchions bearing high-intensity lighting and surveillance cameras and a fifteen-foot-high chain-link fence, and behind that runs a Border Patrol roadway.

Local environmentalists were infuriated about the costly project. They had worked for years to devise a strategy for building a stronger border without destroying the rugged landscape. "We've lost sensitive habitat, and the estuary is now threatened," Jim Peugh, conservation chairman of the San Diego Audubon Society, told the *Los Angeles Times*. "I'm really disappointed that our system wasn't allowed to work the way it has historically and is required to by law." But Peugh's plea to protect the imperiled habitat in the Tijuana River watershed fell on deaf ears. Finessing the requirements of the 1970 National Environmental Policy Act—a Nixonian piece of legislation—came easily to an administration that habitually violated the law or waived regulations. In its drive to stop illegal border crossings, the ends justified the means.

With the burial of the gulches and canyons, one Border agent boasted that it was now "logistically impossible" to sneak across the flattened site. He described the landscape as being "in a lockdown state," an arresting image of what we have lost.

Highway Robbery

John Steinbeck, who in *Grapes of Wrath* championed California's dispossessed migrants, would stand up today for those suffering from another state-sanctioned attempt to rob them of their mobility. Aptly enough, the automobile is at the center of controversy in both eras.

"This is a free country. Fella can go where he wants," an anonymous voice asserts in Steinbeck's Pulitzer Prize–winning novel. Its interlocutor scoffs: "That's what *you* think! Ever hear of the border patrol on the California line? Police from Los Angeles—stopped you bastards, turned you back. . . . Says, got a driver's license? Le's see it."

Cops are a blunt-force trauma everywhere in this novel about a wandering people's quest for their place in the land. Scornful of the Joads and other desperate folks seeking respite in the Golden State from the dust-choked Plains, the "bulls" terrorize uprooted "Okies," whose 1,200-mile odyssey takes place along Route 66, the "Main Street of America." Eden, they discover, is under armed guard.

It feels just as embattled today. Across Southern California police departments are getting their kicks on

Route 66, and a lot of other roads, the old-fashioned way. They're harassing a new generation of the migratory poor in search of a better life, rousting them from their cars, draining their bank accounts with huge fines, and disrupting their lives.

Especially active are those communities the iconic Route 66, known locally as Foothill Boulevard, rolls through. Nearly every weekend evening in summer 2010, from college-town Claremont to old-money Pasadena, a "sobriety checkpoint" has been mounted along the busy arterial that hugs the southern slope of the San Gabriel Mountains. An intimidating cadre of well-armed police, stationed at these brightly lit barricades, has stopped thousands of cars each night. They ask drivers two questions: Do you have a valid driver's license? Have you been drinking? Answer correctly, and you can get waved through. Hesitate or fail to produce the requisite identification, or appear to have thrown back a few, or speak Spanish, and all hell can break loose. Drunk? You'll be arrested and your car towed to an impound lot. Lack a proper license? You'll be cited and your car hauled away. In theory this is a straightforward process, for who can defend drunk driving? Who favors unlicensed drivers?

Checkpoints would seem to be color-blind, but they're not, as my wife and I learned when we drove along Foothill late one Friday night in July. Our first clue appeared in the reaction of the pair of officers who stopped us. After they assessed us under the high-intensity lights, our complexion led to a subtle choreography; the Latino policeman stepped back so his white colleague could approach.

There was nothing subtle about those who were held back (we were not). Clustered along the sidewalk or in groups in adjacent parking lots were scores of Hispanics. Some were on cell phones, presumably calling for rides; others, weary and worn out, sat still. Their vehicles, however, were on the move. Tow trucks were busy pulling cars and pickups to a distant impound facility, red lights glowing in the night.

Their drivers faced more than an evening of inconvenience. If their license had expired or they were driving without one, their car would be held under lock and key for thirty days. (If charged with DUI, by contrast, you can reclaim your car the next day.) This month-long loss of transportation has a devastating impact on those who commute to work by car, and who doesn't in Los Angeles? Magnifying this blow to a poor family's economic prospects are the punitive fees towing companies and police departments charge to release an impounded automobile. In Latino-dominated Baldwin Park, the per car tab runs to nearly $1,500. Storage fees make up the bulk ($1,350), and the rest comes from the city's vehicle release fee. When families cannot pay the city gets a cut of the abandoned car's sale at auction.

This bit of highway robbery has become big business for the many recession-wracked communities of Southern California. At its early August checkpoint, Baldwin Park officers snagged 150 cars in one evening, netting the city a whopping $38,400. Its take for the past fiscal year amounted to $338,000, a healthy infusion of cash for a town struggling to make ends meet.

Pomona, the fifth-largest city in Los Angeles County, boasting a majority-minority population of 160,000, has also cashed in. Like its peers, Pomona secures U.S. Department of Transportation dollars, passed through California's Office of Traffic Safety, to underwrite a number of high-profile checkpoints. Targeting working-class neighborhoods and even Cinco de Mayo celebrations, it has swept thousands of vehicles off the streets. Councilwoman Cristina Carrizosa has denounced the police department's "Gestapo"-like actions, alleging that in 2007 and 2008 local checkpoints boosted city coffers by $1 million.

That kind of return has sparked a gold rush. In 2009, according to the nonprofit investigate team California Watch, checkpoints throughout the state raked in $40 million in fees and fines. Police overtime pay, underwritten with federal dollars, amounted to an additional $30 million. Fleecing the migrants and the poor is big business.

Profiling drivers, however, is of questionable legality. Although checkpoints' stated purpose is to deter drunk driving, California Watch and other media investigations have revealed that only a fraction of the cars hauled away are the result of a DUI. At a checkpoint on Foothill Boulevard in Claremont, 1,570 cars were stopped; only two impoundments were alcohol related. Instead police were stripping cars from unlicensed drivers, the preponderance of whom are Latino. The legal justification for such seizures, the 9th Circuit Court of Appeals warned in *Miranda v. City of Cornelius* (2005), is tenuous.

No surprise, then, that the egregious racial profiling, outlandish financial kickbacks, and questionable legality of checkpoints have sparked heated protests. An acquaintance in the Texas Rio Grande Valley, where stoppages are frequent, refuses to answer officers' questions, passive resistance designed to snarl the operation. Grassroots organizers nationwide steer cars away from checkpoints by standing some blocks away with placards reading "¡Retén adelante!" (checkpoint ahead). Angry citizens in Southern California have jammed city council meetings to challenge their political legitimacy. In Pomona the chief of police was forced out of office. In Baldwin Park, after the lucrative August checkpoint brought 300 enraged voters to city hall, officials quickly suspended their use and released impounded vehicles.

This activism may seem small-scale, but it has the potential to shift the larger political dynamic. In *The Grapes of Wrath,* John Steinbeck noted that face-to-face resistance turned the tide for this earlier set of harassed, car-driving migrants. The "little screaming fact that sounds through all history," Steinbeck wrote, is that "repression works only to strengthen and knit the repressed."

Homeland Insecurity

The U.S. Border Patrol has been under the gun. Beginning in 2004 right-wing politicians began rebuking the agency for failing to secure the nation's borders. Congressmen Tom Tancredo (R-CO) and Lamar Smith (R-TX) were among those demanding a tougher border response and a more rigid law enforcement. Militia groups took up arms to defend American territorial sovereignty, claiming that lawyers and legislators had handcuffed the Border Patrol. And immigration rights activists have gone to court and into the streets to protest the Border Patrol's rough handling of undocumented migrants and its armed raids that disrupt workplaces and households. The agency's shiny badge has been badly tarnished.

In February 2009 it earned another black mark. That month an embarrassed Border Patrol ordered an internal investigation of allegations that some of its Southern California officers were given strict arrest quotas for undocumented migrants. If they did not meet them, officers would be moved to less desirable shifts and locales, and their careers would suffer. Jeffrey Calhoon, chief patrol agent for the El Centro Border Patrol office, ordered the investigation, vowing, "If there is some threatening behavior, we're not going to tolerate it."

But Calhoon also admitted to the Associated Press that the agency regularly sparks competition between patrol units by challenging them to meet arrest figures during a particular month. In late 2007 officers at the Riverside station were given a target of 100 arrests for November and December. In January supervisors raised the quota to 150. That's when some officers went public with their concerns, forcing Calhoon to launch a probe of what appeared to be a time-honored practice.

Why the sudden desire to bolster the number of arrests? No doubt the collapse of the U.S. economy was in good part to blame. The slowdown in construction and related landscape businesses and the sharp decline in retail sales, which forced garment factories and sweatshops to cut workers, left little financial incentive for unskilled labor to cross into the United States in hopes of securing better prospects.

This led to fewer arrests along the border, a fact the agency readily admits but then spins to its advantage, arguing that its agents captured fewer undocumented migrants because of the Border Patrol's increased presence. Yet that success, if that is in fact what it was, also led the agency to leave its primary field of operations and press into the interior of California, hunting for those undocumented who were already established in the United States. One who suspected this strategy was Emilio Amaya, executive director of the San Bernardino Community Service Center. He told the Associated Press that beginning in winter 2009 the Border Patrol had become "very visible" in the Inland Empire region of the state. "This was not an issue in the past," he said.

Its increased visibility emboldened pro-immigrant forces. On February 7, a cool and rainy Saturday, more than 300 people marched from Riverside City Hall to the local Border Patrol office to protest the alleged quota system and denounce agents sweeping through a nearby barrio. "They have targeted the Casa Blanca neighborhood, specifically the area where workers congregate in front of the Home Depot," Omar Leon of the National Day Laborer Organizing Network told the crowd. "Rather than focusing on real criminals they end up targeting workers and their families just to meet their quotas."

For all its outrage, Leon's impassioned plea contained a mournful note. There is something terribly sad about a public policy designed to bolster "homeland security" that produces just the opposite. Anxious officers, apprehensive laborers, and nervous neighborhoods make for an unsettled nation.

Why Friendship Park Mattered

Friendship Park is no more. In January 2009 it fell victim to the bulldozer's blade, another venerable landscape flattened in advance of the border wall's construction.

There is considerable irony in Friendship Park's devastation. Built in 1971 just south of San Diego and straddling the U.S.-Mexico border, it commemorates the launch of the 1849 survey that established the two countries' new frontier in the aftermath of the Mexican-American war. Two California Republicans, President Richard Nixon and California Governor Ronald Reagan, cooperated in the park's development, especially the half-acre plaza known as Monument Mesa that contains an obelisk dedicated to the binational survey crew. To heighten the park's import, the president sent a special envoy—his wife, Pat—to represent him at the opening ceremonies.

The First Lady's presence was not just ceremonial. A photograph shows her working the rope line, which in this case meant walking along the thin barbed-wire fence separating the two countries. As one Mexican citizen lifted the top strand above his head, she reached out to shake another's hand. How perfect that this man cradled a child in his other arm, a peaceful moment Mrs. Nixon underscored when she reportedly declared, "I hate to see a fence anywhere."

Her words came back to haunt us as heavy machines leveled portions of the park so that the Department of Homeland Security could construct a fifteen-foot-high border wall, the final three-and-a-half-mile-long segment separating San Diego and Tijuana. With completion later that spring, the Bush Administration formally interred what was once a conservative Republican commitment to international comity.

Times have changed, of course. Indeed, the Clinton Administration initially reinforced the fence dividing Friendship Park. Yet its reconstruction did not curtail the weekly religious services held there, during which attendees shared communion through the fence. It did not stop families from celebrating life moments with cross-border kin. Nor did it prevent lovers from playing Romeo and Juliet, plighting their troth, though separated by chain-link.

None of these informal, communal, and affective connections is possible any longer. The rationalizations for why it was essential in 2009 to sever these close-knit relations at the remarkable venue hardened like the thicker boundary itself. Through his press secretary, U.S. Rep. Brian Bilbray (R-Solana Beach) copped an uncompromising attitude. "Seems like everyone is looking for an excuse not to build the (new) fence," his spokesman Kurt

Bardella asserted. "National security is something that should never be compromised."

Mike Fisher, head of the Border Patrol office in San Diego, was slightly less severe. "I fully respect what goes on in that park," he said. "It's unique because it is somewhat open. Unfortunately, the smugglers exploit that as well." How convenient that Fisher used their behavior to undercut the family-centered image of the park and advocate its destruction. "We have documented cases where people are selling false documents through the fence," he added. "They are selling drugs through the fence. They are smuggling people and babies through the area." Tough times apparently called for tougher measures.

Did they, though? Without denying that the borderlands have become more dangerous over the past decade or dismissing a nation's sovereign right—and need—to control its territorial limits, there remains real value in preserving points of contact, places of openness and accessibility that daily remind us of our cherished democratic ideals.

If our policy consists solely of building up impenetrable barriers patrolled by heavily armed guards, if it amounts to drawing a very hard line in the sand, we will have violated the real legacy of Monument Mesa. That elevated site, after all, memorializes not just the 1849 survey party's first step in redefining the postwar national boundaries but also the willingness of Mexico and the

United States, despite their intense grievances, to repair a bloodied relationship. The fortified wall that cleaved Friendship Park mocks this invaluable lesson in hard-won reconciliation.

Pat Nixon embodied this lesson when she extended her hand at the border's edge. With one simple gesture she acknowledged that a fence, as substance and symbol, is antidemocratic. We should have followed her generous lead.

Praise Song

At Friendship Park they broke out in song. On February 21, 2009, hundreds of Americans and Mexicans gathered at the embattled site that lies astride the Mexico-U.S. border between San Diego and Tijuana. They protested through music the Department of Homeland Security's construction of an impenetrable barrier sealing off this historic and beloved spatial link between two countries.

The piece they belted out—in voices trained and not—was Gabriel Fauré's Requiem in D Minor. The composition's haunting and evocative melody, like its innovative liturgy, was oddly suited to the task of political resistance that drew so many to the border's edge.

Composed between 1887 and 1890, Fauré's work broke with the requiem's structure that earlier composers such as Mozart, Haydn, and Brahms had employed. He stripped it of the requisite ode to God's awesome power, to a thunderous Judgment Day. "I instinctively sought to escape from what is thought right and proper, after all the years of accompanying burial services on the organ," he told an interviewer. "I wanted to write something different."

That rebellious spirit guided those who organized the cross-cultural chorus of voices opposed to armoring the border. Under the provocative banner—"A gathering of the people, by the people, and for the people of the U.S.-Mexico border region"—the rally drew on a cross-border coalition of faith-based communities, university choirs, and grassroots foundations and organizations.

Their collective ambition had as much to do with politics as with piety. As the Department of Homeland Security shut off access to the park, which was created in 1971, the protestors were determined to insert themselves back into the contested space as a deliberate challenge: "We will be staking our claim to this historic meeting place in the name of the peoples of the border region." The outcry was also intended to have national ramifications, the organizers affirmed. They called on "the Obama Administration to bring an end to the construction of walls on the U.S.-Mexico border, to launch a comprehensive review of plans for border, enforcement and security, and to restore routine public access to Friendship Park."

As American citizens walked toward the border, a wall of Border Patrol officers lined up to stop their progress. From forty-five feet away the singers launched into Fauré's majestic work, singing in concert with a recorded version that swelled up from a sound system located on the other side. "The choir performed admirably, despite having to compete with whistles, shouts, and bullhorn blasts from a small group of anti-immigrant protestors who tried to hijack the gathering," the Rev. John Fanestil, one of the event's organizers, later blogged. "Their inimitable combination of ignorance, hatred, and incivility was no match for the choir, which included a stunning soprano solo—the *Pie Jesu*, 'at the feet of Jesus'—sung from a distance in Tijuana." Armed Border Patrol agents forced the activist minister off the beach when, as he had done for months, he sought to offer communion to those assembled on the fence's southern side. "Go to Tijuana if you want to serve communion," a guard shouted. "You're supposed to be a man of God. Then obey the law!"

Yet Fanestil and those who joined with him were obeying the law when they sang truth to power, albeit a power more supreme than the one the brown-uniformed border agents embodied. Theirs was in pointed defiance of earthly authority's capacity to constrain human experience. This is why their choice of Fauré's Requiem was so inspired. Its words are designed to elevate us above the quotidian, to take us out of this moment in time. It asserts that there is a force more powerful than this mortal coil. "May the angels lead you into paradise," the Requiem's final stanza begins; "may the martyrs receive you in your coming, and may they guide you into the holy city, Jerusalem."

Here there may be no perfect place, as Fauré suggests, no heaven on earth. Still, something divine happened as the binational chorus intoned his words, as their bodies reverberated with sound, their voices rising in harmony. The music floated over the federal agents, slipping past the fence's physical reality, a forbidding structure of concrete, rebar, and chain-link. In that instant the border disappeared.

Political Agency

Margaret Mead's nostrum for democratic activism—"Never doubt that a small group of thoughtful, committed citizens can change the world. Indeed, it's the only thing that ever has"—always seemed too pat, precious, sanctimonious even. The fact that her words are plastered on almost every grassroots organization's website and are quoted with a kind of hushed reverence rings a false note in this age of the globalization of capital and power. Then I watched Barrio de Colores, a tiny makeshift Laredo-based activist group, rock the mighty Border Patrol back on its heels and realized that Mead may have been on to something.

The clash revolved around the thick stands of Carrizo cane that grow along the Rio Grande in and around Laredo, Texas. The Border Patrol had proposed to eradicate a one-and-a-half-mile stretch of the tall tangle—the fast-growing plant can soar more than thirty feet—through aerial spraying. Helicopters armed with the herbicide imazapyr were scheduled to lay down the toxin in late March 2009.

They never got airborne. Nearby residents quickly

formed Barrio de Colores and filed a federal lawsuit alleging that the federal agency had ignored key environmental regulations. Strikingly, the agency stood down, agreeing to remove the cane through mechanical means and by hand-painting stumps with imazapyr. The plaintiffs had every right to be proud of their achievement, so quickly gained.

Theirs has been an all-too-rare accomplishment. Since 2006 other grassroots organizations have challenged the Bush Administration's decision to harden the border by any means possible. From Texas to California, they have rallied against bulldozers scraping wide swaths through urban neighborhoods, rural ranchlands, public parks, and wilderness preserves; opposed backhoes digging trenches so that construction crews could erect chain-link fences and concrete walls; and tried to halt workers from stringing barbed wire, fitting high-intensity lighting, burying ground sensors, or mounting high-tech surveillance cameras atop the infrastructure. To no avail. The unchecked federal power along the borderlands is captured in the chilling photographs of this fortified landscape. These stark images evoke nothing so much as the Berlin Wall. To complete the Cold War–like montage of a brutalized terrain and an abject population, scan the sky for helicopters. Some swoop along the border at night and with powerful infrared searchlights paint those who try to cross into the United States under cover of darkness.

So convinced was the Border Patrol of its decision,

so compliant were local politicians who bowed before its authority, that it did not think twice about the consequences of launching an air armada in a scorched-earth campaign against the nonnative Carrizo cane. As for the broad-spectrum herbicide, imazapyr seemed apt in part because of the martial monikers it is marketed under—"Arsenal" and "Assault."

By all accounts imazapyr is deadly. It attacks foliage and roots, interfering with DNA synthesis and cell growth. It is also an indiscriminate killer. Although the Bush-controlled EPA reapproved imazapyr in 2006, the agency conceded that it is particularly lethal to rare and endangered species, and that while the herbicide is not carcinogenic and has no known impact on human reproduction—an affirmation the Border Patrol used to justify employing it—direct contact may result in rashes, swelling, and other irritations.

The Critical Habitat Project has offered a blunter assessment of the chemical compound's possible effects. "One primary breakdown product of imazapyr," it reports, "is quinolinic acid, which is a neurotoxin and can cause symptoms similar to those in Huntington's chorea such as loss of coordination and trembling." Such troubling information about imazapyr propelled those living in Lardeo and Nueva Laredo to mount a highly public cross-border protest. It also undercut the Border Patrol's blustery rebuttal of the agency's critics. "We're not out to 'poison' anybody. I find that word a little bit over the top," spokesman Chuck Prichard said. "We're just the guy that needed his house painted . . . This is the method they chose. Based on what we know this would be an effective, low-risk way to do this."

House painting—if only the Border Patrol's plans had been so benign. And if they had been, why not hold hearings about its intended actions or issue warnings to those living in Los Dos Laredos? Why not file an environmental impact statement about imazapyr? The Border Patrol did not take these logical steps because it felt free to ignore federal law, endanger public health, and pollute the earth in the name of what it believed to be the greater good.

Happily, its self-declared freedom of action was curtailed in this case—thanks to that small band of activists, Barrio de Colores.

Bulldozing Nature

The past has a funny way of maintaining a tight grip on the present. Ask anyone who cares deeply about the pristine remnants of Southern California's unique arid landscape. In late winter 2009 bulldozers began to rumble through the Otay Mountain Wilderness Area, a sanctuary in eastern San Diego County, prepping the ground for the border wall to slice across its desert scrublands, steep canyons, and rugged high ground.

The small reserve—it encompasses only 18,500 acres—was established in 1999 and is managed by the Bureau of Land Management. Despite its limited size, the Otay is of crucial significance. According to the Environmental Protection Agency, San Diego contains the greatest number of threatened or endangered species in the continental United States; its high desert is home to many of them, including the quino checkerspot butterfly, the arroyo toad, and the Otay Mesa spreading mint. This often bone-dry terrain is also vital for migratory mammals, whose search for food, water, and shelter know nothing of national boundaries.

This viable habitat and the rich biodiversity it has

sustained for millennia came under attack in December 2008 when Michael Chertoff, secretary of the Department of Homeland Security, waived the Wilderness Act and a host of other protective legislation so that a contractor could scrape clean this untrammeled terrain. Site preparation for the steel wall's construction, which included constructing the requisite hardened roadway, was completed a year later.

The impact was immediate and pronounced. Nathan Trotter, a local activist who toured the area in January 2009, noted that the initial cuts for the road and wall were so great "that the resources needed to restore the area would be immense." Other observers reported that the construction activity had accelerated erosion and washouts.

None of this destruction needed to occur. The Border Patrol itself did not think the wall was necessary. Richard Kite, a spokesman for the agency's San Diego office, told reporters in 2006 after Congress passed the Safe Fence Act, "It's such harsh terrain it's difficult to walk, let alone drive. There's no reason to disrupt the land when the land itself is a physical barrier." The EPA also cast doubt on the project. In a February 2008 letter to U.S. Customs and Border Protection, it objected to "the filling of two well developed riparian corridors in Copper and Buttewig Canyons and [had] concerns regarding high potential for significantly increasing erosion in the watershed from the combination of road widening, new vehicle trail construction, fence installation on steep slopes, and fence installation across intermittent streams." The EPA predicted that these intrusions would "have unacceptable adverse impacts under Section 404 of the Clean Water Act, especially considering cumulative impacts from other border fence projects that are proposed in the Tijuana River Watershed. These impacts must be avoided to provide adequate protection for the environment."

Wilderness advocates were much more outspoken about the Bush Administration's decision to savage the Otay, including blasting canyon walls, trucking out more than 500,000 cubic yards of fill, and grading and leveling a 150-foot-wide swath on which to erect an expensive wall (more than $16 million a mile). Carl Pope, executive director of the Sierra Club, said, "Wilderness areas are designated by Congress specifically to protect sensitive places from projects like this road construction. This road sets terrible precedent and clearly demonstrates the dangers of granting the Secretary of Homeland Security authority to waive any law in order to build walls along our international borders." Matt Clark, Southwest representative for Defenders of Wildlife, said, "Such harmful impacts to wilderness characteristics and values are clearly inconsistent with the congressional intent of the law that established the Otay Mountain Wilderness Area in 1999. The waiver and the wall are an affront to our nation's laws and natural heritage."

The Bush Administration ignored these principled

arguments, and the Otay Mountain Wilderness Area, like countless other sensitive ecosystems along the U.S.-Mexico border, paid a heavy price. We have paid, too. By compromising the Otay's historic function as wildlands, preventing resident animal populations from taking full advantage of their primeval habitat, and subverting national environmental regulations, the border wall casts a long shadow over contemporary American politics, a grim and costly legacy.

Behind Bars

I have never seen an ocelot, never spotted a jaguarundi. I have never stalked either of these furtive wildcats in their native habitat in South Texas, the rare sabal palm forests that once grew in thick profusion along the slow-moving Rio Grande near its confluence with the Gulf of Mexico.

If the Department of Homeland Security has its way, I may never catch a glimpse of these imperiled animals in the junglelike ecosystem that has sustained them for millennia. In December 2008 the department announced that it would build a section of the border wall through the Lennox Foundation Southmost Preserve, a private sanctuary the Nature Conservancy manages. The conservation organization challenged the federal agency's right to do so, and the DHS went to the U.S. District Court in Brownsville, seeking its approval to use eminent domain so that it could condemn a broad strip through the protected landscape. At stake was its right to construct a security fence along a corridor 60 feet wide and 6,000 feet long. As compensation for the fence work, DHS offered the conservancy the magnificent sum of $114,000. The dispute remains in court.

In rejecting the federal government's willingness to run roughshod over communities and landscapes, the Nature Conservancy joined with other environmental and grassroots organizations struggling against the border wall that has bisected the Rio Grande Valley. Even before it went to court, the Nature Conservancy had signaled its disagreement with the wall as the sole solution to border security. In a September 2007 press release it laid out its opposition on environmental grounds. Because the 1,250-mile-long Rio Grande Valley is so fertile and its "lushly vegetated river banks provide an irreplaceable wildlife corridor that resident and migratory animals and birds depend upon," and because "the spectacular diversity of wildlife in the Valley has led to recognition of the area as a major ecotourism destination and is a fundamental economic component, contributing $125 million a year to the local economy from nature tourism," the border wall would jeopardize natural and built landscapes.

These concerns were particularly pressing, the conservancy stated, because the valley's staggering growth over the past two decades had fragmented and shrunk open space. Thus the Southmost Preserve, situated on more than 1,000 acres in Cameron County, like the small federal wildlife refuges scattered up and down the borderlands, was of even greater significance to preserving biological diversity and sustaining ecotourism.

Two years later the Nature Conservancy maintained its commitment to these ecological demands and economic imperatives and rejected the DHS's effort to condemn and compensate. "The financial offer we received from the federal government is shortsighted and only takes into account the footprint of the border fence itself," said Laura Huffman, the Texas state director of the conservancy. "It doesn't begin to make up for our inability to manage the more than 700 acres of our preserve that lie between the proposed fence and the Mexican border. This land is, quite literally, irreplaceable."

These acres, known as the "jewel of the Rio Grande Valley," support several landscape restoration projects in which researchers are regenerating indigenous palms and other plants that were plowed under for agricultural operations or flattened by urbanization. They are also assessing threatened and endangered flora and fauna, including the Tamaulipan thorn brush and indigo snakes, Texas tortoises, and local rarities like the Brownsville subspecies of the common yellowthroat. On the preserve's grounds the conservancy educates students and citizens about these wildlife communities and their evolution and import. Communicating this knowledge to other threatened terrain is vital to the conservancy's stewardship mission.

Its vital outreach programs, however, will be meaningless if it is unable to protect the remaining acres under its management. That's key to understanding why the conservancy has been so opposed to the DHS lawsuit and the eighteen-foot-high steel and concrete fence that would sever the rare habitat under its care. It knows, too, that by disrupting the migratory patterns of the ocelot and jaguarundi, the border wall would convert the preserve into a prison.

Walled Off

Pygmy owls and bighorn sheep make an odd couple. These keystone species along the U.S.-Mexico border have little in common; their size and weight (so perfectly captured in their names), like their mode of transportation and food sources, could not be more different. Yet they share a desert habitat, and therein lies their mutual problem.

The impenetrable border wall the United States is building across the southwestern states will inhibit their ability to move, hunt, and reproduce. How ironic that this triple-thick bulwark, which has failed to stem the flow of undocumented migrants into the United States (to do that required the collapse of the global economy), may have a fatal impact on animals innocent of the human politics and policies that led to its erection. Two conservation biologists, Aaron Flesch at the University of Arizona and Clinton Epps of Oregon State, have been exploring the troubling possibilities the wall may pose to these two species' daily life and long-term survival.

Consider the pygmy owl's below-the-radar flight path. Most of its sorties are conducted less than thirteen feet off the ground, as it swoops down on its prey from its preferred perch in a secluded thicket. A wall that stands more than nineteen feet and whose construction requires bulldozing a wide strip on either side thus poses serious problems for individual owls to protect themselves from predators and sustain themselves over time. To complicate things further, juvenile owls need undisturbed landscapes to colonize in, a nesting behavior the wall's structure necessarily will frustrate. And the breeding colonies on the American side of the border apparently require contact with those on the other. As Flesch observed, "Movement of pygmy owls from Mexico to Arizona may be necessary for the persistence of the Arizona population."

Cross-border fertilization may prove even more important for the bighorn sheep's survival, Flesch and Epps argue. Their research indicates a strong genetic link between sheep populations living in the Sonoran mountains of Mexico and those inhabiting the southern highlands of Arizona. Without unimpeded connectivity the different herds will become cut off, degrading their genetic diversity. "Bighorns in places like the Sonoran desert will form small populations—sometimes with only ten, fifteen, or twenty animals," Epps said in a press release. "Yet they will occasionally move back and forth and mix with other groups. That connectivity is critical to their survival. Without it, they can still sometimes recolonize, but often that small group will go extinct." Should that occur, these sheep, like the pygmy owl, will find that human agency has sabotaged their biological imperatives.

The political impulse that built the wall in defense of U.S. borders is itself a form of territorial marking, a behavior every species manifests. All species try to manage their habitats to enhance their chance of survival—a process, often enough, that depends on beneficial interactions with other creatures; this mutuality is critical to the creation of sustainable landscapes and the rich biodiversity that makes them so.

But *Homo sapiens* also loves monocultures, as our suburban lawns, big-box malls, and agricultural practices testify. Through our tools and technology we have the power to clear away competitors or build environments that diminish their odds of survival. The border wall is a monument to human hegemony. Yet it is only one expression of our dominance. As contractors flatten the land, dig fence posts and hang chain-link and concertina wire between them, set up guard towers complete with searchlights and infrared cameras, and lay down hardened roadways for high-speed patrol vehicles, they are manufacturing a landscape with a single purpose. The goal of this high-tech infrastructure on the ground, when combined with aerial surveillance, is to halt illegal immigration into the United States. The federal government has spent tens of millions of dollars, trained thousands of new border guards, and flexed its fiscal and physical muscles, all to display its authority.

One manifestation of this is reflected in the commotion Epps describes as integral to the new border security habitat. "The sheer number of people moving across the landscape is stunning," he says. "The crackdown on urban crossings has forced people into wild areas," intensifying "the amount of human activity—both people crossing and the border patrols seeking them" in an arid region that was once thinly settled and rarely visited.

Bearing the brunt of this escalating human presence are the wildlife Flesch and Epps have been studying. They propose a number of changes to the border wall's structure to increase the chance that pygmy owls and bighorn sheep will survive in this trampled terrain. For the owl, they suggest maintaining tree cover along the border, and for the sheep, a more open wall that will allow them to continue their cross-border migrations. Even so, they recognize that their proposals conflict with the governmental desire to maintain an impassable barrier and are thus nonstarters.

Indisputable is their larger, unstated argument: to worry about the survival of wildlife, regardless of size, is to be anxious about our own.

Just Litter

In December 2008 Walt Staton spotted a U.S. government helicopter tracking him as he hiked in the Buenos Aires National Wildlife Refuge in southern Arizona. He suspected he was in trouble, and he was right.

When he returned to the trailhead, federal agents were waiting, and they promptly busted him for "premeditated littering." It was an odd charge. A member of the grassroots organization No More Deaths/No Más Muertes for several years, Staton has been leaving one-gallon water bottles along the paths undocumented migrants have created in the rugged mountains southwest of Tucson. The group's goal is humanitarian and not political—to ensure that no one dies of thirst.

Apparently preferring dead bodies to litter, the U.S. Fish and Wildlife Service, which manages the refuge, ticketed Staton for packing out empty jugs. "We always clean up after ourselves," he told the *Claremont Courier*. "So it's ironic that we got tickets while carrying in trash."

There is more than one irony to this story. Fish and Wildlife Service rangers had been aware Staton and his peers' actions since 2004. Although the agency could not condone the actions of No More Deaths/No Más Muertes, it chose to ignore them until 2008, presumably because it was also concerned about the safety of people trying to cross the borderlands. There is no precise tally of those who have died during these difficult passages, but estimates range upwards of 6,000 since the mid-1990s.

Staton, a graduate student at the Claremont School of Theology, argues that the number began to rise after the 1994 enactment of the North American Free Trade Agreement, in response to which "the United States has stepped up its border security in anticipation of a worsening of the Mexican economy and a rise in illegal immigration." His analysis is accurate as far as it goes, but it is also true that the major pressure to clamp down on border crossings occurred after 9/11. The Republican Party politicized the issue with the aid of right-wing commentators such as Lou Dobbs and Rush Limbaugh. In Congress, Rep. Tom Tancredo and Rep. Lamar Smith, among others, pushed to increase the Border Patrol's budget, firepower, and authority; deputize local police forces to arrest the undocumented; and construct what they hoped would be an impregnable border wall. Together these punitive initiatives funneled potential migrants away from border cities and into the uninhabited and dangerous deserts of southeastern Arizona. Those who died in this tough terrain did so because of American politics.

Staton's arrest calls into question this controversial history and devastating public policy. In August 2009 a federal jury sentenced him to 300 hours of community service, one year of probation, and a one-year ban from the Buenos Aires refuge, punishment that conservative blogger Debbie Schlussel thought too lenient. "In my view, the guy should have gotten jail time for helping along the alien invasion of America, and he should have received a ban from the refuge for far more than a year, perhaps a lifetime," she wrote. "We need to put our foot down against these liberal illegal alien activists who commit what is tantamount to treason."

For his part, Staton challenged the court's actions, filing an appeal that was heard that December. He lost and subsequently agreed to do the required hours of community service. He vowed not to comply with the ban from the refuge, however. "We shouldn't accept the punishment . . . for doing humanitarian work," he said. "We will not be deterred."

Wandering in the Wilderness

Republicans and some conservative Democrats in the West have been gunning for the Endangered Species Act since its initial iteration was enacted in 1968. They have been targeting the 1964 Wilderness Act even longer. They and their comrades-in-arms nationwide have been sniping at immigrants with mounting ferocity since 9/11, and their attacks reached a fever pitch in the last election cycle. Imagine the fury that would erupt if the GOP could combine these seemingly unrelated animosities in one place—the U.S.-Mexico border, a landscape they love to loathe.

Rep. Rob Bishop (R-UT) tried to fuse these explosive issues as soon as he became chair of the House subcommittee on public lands in January 2011. "It is unacceptable that our federal lands continue to serve as drug trafficking and human smuggling superhighways," he declared. Along the 1,933-mile southwestern border, "strict environmental regulations are enabling a culture of unprecedented lawlessness." To combat these incursions, Bishop, whose Ogden, Utah, district office is closer to Canada than it is to Mexico, proposed a regulation allowing Border Patrol agents free movement across all public lands, including those designated as wilderness. Its enactment would resolve what he dubs a "conflict between wilderness and border security."

Whether such conflict existed is doubtful. In October 2010, for example, the Government Accountability Office noted little tension between the Border Patrol and federal environmental regulatory agencies. That finding did not stop the *Wall Street Journal* from extolling Bishop's posturing on border security issues and his efforts to ignite another Sagebrush Rebellion, a longstanding western pushback against federal management of public lands that dates to the founding of our remarkable system of national forests and parks in the early twentieth century. One hundred years later figures like Bishop are still trying to undercut these national treasures and the pristine landscapes, scenic vistas, and cultural values they embody.

Adding a xenophobic twist to this studied antifederalism has already produced disturbing results. Nowhere is this more obvious than in Southern California's Otay Mountain Wilderness. The 16,893-acre site, a rough high country near the border east of San Diego, is home to endangered butterflies, relict stands of cypress, and other threatened flora and fauna. Although the Bureau of Land Management manages the area under the provisions of the Wilderness Act, in 2005 and 2006 Republican-controlled Congress stripped its authority to protect the

Otay's unique features through the REAL ID Act and the Secure Fence Act. These allowed the Department of Homeland Security to waive every law in its way, from endangered species to clean water to noise control, so that it could construct the controversial border wall. In the Otay this entailed bulldozers crashing across breathtaking mountainous desert, blasting fragile habitat, obliterating steep canyons, and disrupting wildlife migration. Similar devastation occurred in refuges, preserves, and other wilderness areas in Arizona and along the Rio Grande Valley and in Texas.

Only time will tell whether Republicans can shred our country's environmental protections by linking them to popular anxieties about illegal immigration and federal authority. That the GOP has chosen this strategy to increase its power in the run-up to the 2012 elections is a sign of how valuable the borderlands are as political dynamite, and how vulnerable they are as a place in the American imagination.

FOUR Southland

On Fire

It's never a pretty sight when Southern California burns. Images of enflamed ridgelines, smoldering churches, gutted Jaguars, and charred homes, along with an arsenal of firefighting equipment—especially the ubiquitous, low-flying fixed-wing air tankers and helitankers—and yellow-coated, grime-stained firefighters cutting firebreaks, are the staple of the nightly news every summer and fall during the region's fire season.

Once in a while the images fail to capture the destruction's all-encompassing immensity. That was the case in 2007. The scale was astonishing (fires erupted along a 200-mile-long stretch from Santa Barbara to northern Baja, Mexico); the costs were overwhelming (San Diego alone suffered damages topping $1 billion); the loss was mind-boggling (hundreds of thousands of people were displaced, and 500,000 acres and 1,700 homes were burned). And how to convey the smell of fire? During the monstrous Station Fire in 2009, the largest blaze in Los Angeles history—more than 250 square miles were burned—the acrid stench was inescapable; the air tasted like charcoal. Some days the smoke was so thick that the sun was blotted out, and day became night.

Most confounding has been the human response. Although no one questions that Southern California has always burned, each year anguished homeowners are convinced life will go back to normal once these particular fires are extinguished. Their conviction is understandable, the desire normal. Still, it's the last thing that should happen, for these fires warn of devastating fires to come.

We can get a glimpse of this fiery future by paying attention to the recent past. The massive October 2003 burn, for example, killed sixteen people and torched over 600,000 acres; those in Riverside County in 2005, although not as extensive, were deadly for firefighters. As big and dangerous as these fires were, those ignited between 2007 and 2009 burned hotter, with a greater intensity, for a longer period of time, and at a much greater cost than their immediate predecessors.

What can we do? Start by recognizing that Southern California is a landscape fire helped construct. Like other Mediterranean climes around the world, it has always gone up in smoke. Many of its ecosystems are fire-adapted. Resident animal populations have evolved in relation to the fire cycles (and not just in knowing how to run away!). Some tree species, such as the lodgepole pine, can only regenerate as a result of the searing heat these annual infernos can generate.

Early Californians proved every bit as adept. They

reaped the nutritional benefits of fire's aftermath (think mushrooms). As elsewhere, they used fire to open up forests and create meadows, all with an eye to expanding their hunting and gathering prospects. As ethnoecologist M. Kat Anderson has argued in her seminal 2005 book, *Tending the Wild: Native American Knowledge and the Management of California's Natural Resources*, "Many of the natural features of the present-day California landscape are not, strictly speaking, 'natural,' but are rather in part the product of deliberate human action."

These ancient peoples grew up with fire, recognized its power, and knew how to get out of its way. That last lesson seems to have been lost on modern Californians in part because we believe ourselves independent of the natural world. The houses we build, the freeways we travel, the water we pump, even the air we breathe reinforces our belief that we live in a built landscape beyond nature. We imagine ourselves liberated from the restraints it imposed on other people in earlier times.

This self-congratulatory perception gets us into trouble. Our high-speed auto-infrastructure allows us to flow across Southern California without restraint, and although it has made it possible for us to penetrate deep into the canyons and foothills, it has also placed us smack in the middle of historic fire zones. We have built thick clusters of grand homes in areas that are supposed to burn.

Rather than enact public polices that would stop

such construction and thus protect us from ourselves, we commit local, state, and national resources (human and fiscal) to extinguish these fires regardless of cost. The price tag will only increase, thanks to the energy the Industrial Revolution has unleashed over the past 150 years. The more we produce and consume, the greater the impact on global temperatures; their slow elevation has had a demonstrable impact on the length of the fire season, as well as on its intensity and destructiveness. Should this process and the attendant drought continue, an increased number of infernos will constitute the region's "new" normal.

To minimize these looming mega-fires' impact will require a new policy regime. Regional planning authorities and zoning commissions must take more control of development in urban wildfire zones. More than six million homes are located in this imperiled terrain, and their number is expected to grow by 20 percent in the coming decade. Local governments and insurers can stop their construction by modeling fire-control regulations on those adopted along the flood-prone Mississippi River valley, the Gulf Coast, and South Texas. Legislation there prohibits insuring construction in areas known to go underwater, and local governments are purchasing older homes in floodplains, tearing them down, and creating greenbelts along rivers and coastlines.

Localities and the state must also hammer out codes to minimize fire damage to houses and buildings and to enforce the clearing of "defensible space" around the perimeters of individual homes and communities. Finally, the powerful California congressional delegation should continue to push for a long-overdue restructuring of federal funding for fire suppression. Currently the U.S. Forest Service, which manages 193 million acres of forests and grasslands, including the Los Padres, Angeles, San Bernardino, and Cleveland National Forests that ring Southern California, spends 45 percent of its annual budget fighting fires. To make ends meet, it strips funds from its program to reduce hazardous fuels. Forced to spend millions on short-term suppression, the Forest Service is unable to thin ready-to-burn forests. Congress ought to correct this untenable situation by establishing a stand-alone, emergency fund for firefighting, freeing the agency and its partners to better manage these invaluable public lands before they burn up.

If we are unable to live in alignment with the fire ecosystems Southern Californians inhabit, the dark pall that hangs over the Los Angeles basin almost every fall will remain a sign of our habitual failure to understand the landscapes we call home.

Up in Smoke

Blackened skies. Scorched earth. Gnarled oaks lit up like flares. The whomp-whomp-whomp of helicopters aloft. Evacuees huddled in a school. These quick-cut images announced that the 2010 fire season in California had begun.

Although the region's major fires were speedily contained, and although the trio of blazes erupting that July proved to be the only ones of significance all summer and fall, the larger lessons were patently obvious: the stunning devastation from the 2009 Station Fire and the political inferno it ignited drove a much more aggressive firefighting response one year later.

The wind-whipped Bull Fire, which started on July 26 in the Sequoia National Forest in Kern County, within three days swept through 16,000 acres of grass and brush along the Kern River, near the town of Kernville. Nearly 2,400 firefighting personnel had at their command an impressive arsenal of 124 engines, 5 bulldozers, 16 water tenders, and 14 helicopters to battle the blaze. This ground and air technology, combined with the 99 hand crews who did the essential backbreaking labor to clear fire lines around the perimeter, were signs of the seriousness with which this early outbreak was taken.

Every bit as significant was the swift reaction to the West Fire. It flared up on July 27 to Tehachapi's southeast, not far from major wind energy farms that supply Los Angeles with power. The same gusting force that turns those turbines propelled the fire through the kindling-dry landscape, and within two days it had burned an estimated 1,400 acres. However small it appeared, the West also drew a major crew to extinguish it. More than 1,000 firefighters from county, state, and federal agencies worked in conjunction with eight helicopters, nine fixed-wing aircraft, and a fleet of engines and bulldozers.

Gov. Arnold Schwarzenegger made a photo-op, whistle-stop appearance at Tehachapi High School, command post for the West's operations. He did what governors should do—praised the first responders and declared Kern County a disaster area, which freed up additional state funding for the duration of the emergency. But his most critical contribution was just showing up. His presence underscored the argument that politics, not science, was the determining factor in when and how post–Station Fire conflagrations would be fought.

That said, it is also true that another reason so many firefighters raced to the West Fire, and another 1,750 were dispatched to the fast-moving 14,000-acre Crown Fire near Palmdale, was that these terrains had not burned in a long time. The fuel load and thus the level of fire danger were extreme.

Yet another reason so many resources were hurled at the Bull, West, and Crown fires was the searing memory of the 2009 fire that got away. In August a small arson-ignited fire in the Angeles National Forest blew up, and when the Station Fire was brought under control in October it had torched a broad swath of the San Gabriel Mountains, killed two firefighters, and destroyed countless structures, burning more acres than any fire in Los Angeles County's history. It became a direct warning to firefighting agencies at all levels of government. So did the subsequent congressional hearings that charged the Forest Service with mismanagement and the media investigations that unearthed damning evidence about a possible cover-up of its bungled handling of the fire. Because no agency director—in this case, the chief of the Forest Service—wants to endure such intense public scrutiny, and because no one wants to bear witness to the anguish of burned-out communities, every fire after the Station Fire would get hit hard.

This is not always the smartest response, however. Not all fires must be controlled; some are essential to maintain ecosystem health. Not all firefighting makes economic sense either. While the commitment to protect human life is nonnegotiable, Californians and others across the West must become smarter about where they choose to live. If they decide to reside in fire zones, they need to learn how to safely inhabit those areas so as not to endanger the lives of those racing to their rescue.

In the immediate aftermath of the Station Fire, these cautionary insights have been incinerated. Because fire has become so politicized, whenever sparks fly, a small army of firefighters will storm in and flame retardant will rain down. No questions asked.

Sliding Away

Henry David Thoreau never visited Southern California, but the nineteenth-century poet and writer seems to have a better handle on some of our environmental dilemmas than we do. So I concluded on a rainy February weekend as I reread some of his essays. With wind-whipped storms crashing overhead and water and mud churning through arroyos, channels, and culverts—a dark-brown rush that would surge into the white-capped sea—I found what I was looking for.

"I wish to speak a word for Nature," Thoreau declares in the opening lines of "Walking," and "for absolute freedom and wildness, as contrasted with a freedom and culture merely civil." This salvo was aimed at the comfortable of his day, those who had done well but did not live well, who were not conscious of their place in the landscape that sustained them. To these readers he issued his radical challenge—"to regard man as an inhabitant, or a part and parcel of Nature, rather than a member of society." His proposed reorientation, his demand that we think of ourselves not simply as social beings but as natural ones, dependent on an enveloping environment, made good sense in 1862 when his piece appeared in the *Atlantic*, a month after his death. It might make even better sense now, a century and a half later.

Nowhere may we be more in need of Thoreau's wise counsel than Los Angeles. Take the media's breathless accounts of the mudslides that filled backyards, shoved houses off their foundations, and blocked roadways; the snow that iced over and shut down the Grapevine and Cajon Pass through which interstates snake north and east; and the ocean's power as it ground down beaches, ripped through wharves, and smacked into beachfront housing.

The striking images that accompanied these breathless reports and dazed people whose interviews made it to screen or print were of a piece. Whether people had lost their homes or were stranded on a frozen stretch of road, they seemed unaware that their catastrophic circumstances were directly related to where and how they lived. In Thoreau's terms, they thought of themselves only as social creatures, living outside the natural forces that overran their lives.

Angelenos should not have been so clueless. We know that every summer the foothills and mountains burn. We know that every winter driving rain can fall on torched terrain. We know what hillsides of wet unstable soil can do, just as we know that Pacific storms can generate monster surf and deadly waterspouts. No one living in upland communities or coastal towns should be surprised by their vulnerability, yet every tragedy seems fresh. And in the wake of each storm—fire or water-borne—we confirm our desire to carry on without pause or reflection. All that is needed, we tell ourselves, is a more robust firefighting regime, a greater investment in flood-control infrastructure, stronger seawalls. Money and technology will protect us.

These tools' real function is to shore up our illusions. So John McPhee argued in *Control of Nature* (1989), a withering analysis of Los Angeles's budget-busting efforts to suppress fires, which instead sparked more intense conflagrations; and its Sisyphean labors to build debris basins to contain mudslides, whose effectiveness is compromised every time new houses are built higher in the canyonlands. The world we have fabricated is responsible for many of our intense environmental predicaments.

There is another way. But it requires a philosophical shift in attitude, a reconception of ourselves inhabiting nature—fully, consciously. In "Walking," Thoreau offers us this transformative possibility in his blunt assertion "that in Wildness is the preservation of the World." He was not, I imagine, urging us to preserve wilderness separate and apart from human settlement. That was John Muir's conviction, and it is one reason many contemporary Americans have little sense of nature's daily presence in their lives—it is out there, somewhere well beyond the last suburban cul-de-sac, knowable perhaps only on vacation. What Thoreau was getting at is how a resident of his

Concord, Massachusetts, or our Concord, California, can live within nature, become integrated with its cycles and rhythms, be alive to its quotidian presence and occupy a shared universe. He accomplished this by a simple expedient, putting one foot in front of the other. He walked and thought about the act of walking along dusty street, leafy trail, the rugged wild, and back again. In that circular, if circuitous, route he caught how the varied landscapes he passed through dissolved into one another, a dissolution that made them whole, a world preserved.

Shaken and Stirred

I was grading a batch of papers in my academic office on a Sunday afternoon when the earth moved. Suddenly. Sharply. To be precise, it rolled. My desk seemed to lift and tilt; the walls appeared to bend inward. As I leapt up I could not find my footing. I knew what I was supposed to do in an earthquake—crouch under my sturdy desk. Instead I ran outside, which, had this been a major temblor, might have put me in considerable danger from falling objects and fragmenting glass. Fortunately the ground's wavelike turbulence stopped just as I left the building.

Even as I was scrambling out the door, students were running in. The building is home to the college's geology department, and in the lobby sits a large plasma-screen seismograph that relays information about earthquakes around the world. Over the first months of 2010 it was unsettlingly active. In early January I was transfixed as data streamed in from the massive 7.0 quake in Haiti; on one of the screens the map of the battered island republic was covered in a giant pulsating red dot. It expanded and contracted for months thereafter, a vivid reminder of the fault line's continued spasms. That February a similar mark blossomed over Chile after portions of the country were flattened by one the greatest quakes on record, registering 8.8 on the Richter scale. Two months later the screen filled with information about the third major rupture in as many months. Our students knew exactly where to go to learn what had rocked their quiet Easter Sunday.

The young geologists crowded around the monitor, charting each aftershock, discerning their magnitude, and discussing their significance. Some were on their mobile devices, scanning websites for breaking news about the quake's epicenter. When one shouted out that the U.S. Geological Survey, the federal agency that tracks the earth's physical movements, estimated that the earthquake measured 6.9 (revised upward to 7.2), their excited voices fell silent. They knew that somewhere people and communities had suffered a crushing blow.

They were right. Centered south of Mexicali, Mexico, the April 4, 2010, temblor had a profound impact even on a largely agricultural section of the country. Roadways were wrenched apart, commercial buildings buckled, houses collapsed, a portion of a hospital slid off its foundation, and essential utilities were ruptured. Fires were ignited when gas mains broke, and *bomberos* raced to extinguish them. In nearby Calexico, California, blocks of the downtown core, where many structures date back to the 1930s and 1940s, were cordoned off. The high level of damage after the initial shockwave was exacerbated by a swarm of aftershocks, one of which registered 5.1, rattling nerves and taking down more buildings on both sides of the border. "Things have gotten worse," a worried Hildy Carrillo told the *Los Angeles Times* a week after the big shake. "Roofs and walls that were hanging by a thread after Sunday are coming down." As more of them toppled, and as bulldozers and earthmoving equipment arrived to clear the debris, Calexico was altered, losing its look and feel as a historic border town.

Fortunately the casualties resulting from the Baja quake were significantly fewer than those that so traumatized Haiti (more than 200,000) or Chile (nearly 500). Only two people were confirmed dead. Yet despite the relatively small loss of life, it took a long time before the region's most damaged places recovered, physically and psychologically. The social reverberations, like the ongoing aftershocks that for weeks lit up the seismograph outside my office, would be felt for years.

As the adrenaline drained away, as the undergraduates began to drift outdoors, I wondered idly if there might be a possible upside to this otherwise grim story. What if the vibrations from this seismic activity were to crack through Americans' frightened insularity? What if they were to break down our fortress mentality? This quake, after all, gave the students and me a graphic lesson in how geology made national boundaries seem immaterial, insubstantial. The Baja wave raced north along the Laguna Salada fault and shot across the international frontier, mocking the border wall with its thick fences, heavily armed guards, and high-tech ground sensors. Why not take our cue from the path this earthquake's surging energy followed and ditch the Maginot Line–like defenses we have cast in concrete? Why not create instead a diplomacy defined by the tectonic ties that bind our two countries? The unstable earth that lies beneath our feet, Mexican and American alike, could serve as the foundation for a new transnational solidarity.

On the Wild Side

Hiking in the San Gabriel Mountains with Wayne Steinmetz is an exercise in humility. It isn't just the brisk pace the retired chemistry professor sets as he pushes along trails that are dry as dust or slick with snow, though his stamina is intimidating. More striking is his encyclopedic knowledge of this rough landscape's geology and history and its surprising biological diversity; he knows its life and lore. Ask him a question about the forest's shifting composition, and he will unravel its ecological complexity. Lose your way along any number of steep trails that thread up Mount Baldy, and he can give you step-by-step directions for how to regain your path.

Steinmetz's principled commitment to preserving these rugged lands' integrity is one reason I initially hoped against hope that the Angeles and San Bernardino National Forests Protection Act, which Rep. David Dreier (R-San Dimas) introduced in early 2011, would gain congressional approval. If it had passed, it would have expanded the forest's Cucamonga and Sheep Mountain Wilderness Areas by 18,000 acres, protecting them from development and limiting human activity; anglers and backpackers, mountain lions and black bears would have been the big winners.

Naturally this legislation was only a partial fix. Wilderness advocates had pressed for a larger area to more fully preserve unique riparian features, endangered wildlife, and old-growth forests of chaparral, oak, and pine. Yet even this half measure would have gone a long way toward establishing the political context for addition designations to come, no small achievement in Southern California, where four national forests—the Los Padres, Angeles, San Bernardino, and Cleveland—drape the region's mountainous extent.

The legislation was a vital reminder of why these public lands were created in the first place. Following the passage of the 1891 Forest Reserve Act, which granted the president the authority to carve national forests out of the public domain in the West, Presidents Benjamin Harrison, Grover Cleveland, and William McKinley designated over 45 million acres as forest reserves. Most of these lands received greater federal protection to prevent their continued despoliation from intense grazing, logging, and mining.

Such was not the case in the Southland. Its woods were never logged as heavily as those along the Rockies, and sheep, which John Muir decried as "hoofed locusts," never chewed up its grasslands to the extent that they had in Oregon and Utah. The Southern California national forests (all of which were established between 1892 and

1893) had a different purpose—to protect, restore, and manage watersheds, notably the headwaters of the Los Angeles, San Gabriel, and Santa Ana Rivers. Even then, with a significantly smaller population base, downstream interests—conservationists, citrus farmers, and business leaders—recognized how much this semiarid region's prosperity depended on a clear rush of clean water.

Today one-third of Los Angeles County's water flows off the Angeles National Forest, so it is no surprise that its contemporary defenders make the same claim. As one ardent supporter of the Dreier-sponsored act told Californiawild.org, these "unspoiled rivers still need to be permanently preserved for future generations."

This heartfelt sentiment aside, the chance of any wilderness legislation getting through the Republican-controlled House of Representatives was slim. When President Obama directed the Bureau of Land Management to denote more lands as wilderness, a dramatic break from his immediate predecessor's anti-environmental policies, western conservatives went ballistic. The GOP immediately promised congressional investigations into the "radical extremists" who they alleged had the president's ear. Growled Rep. Rob Bishop (R-Utah), "This is little more than an early Christmas present to the far left extremists who oppose the multiple use of our nation's public lands."

Tempting though it was to dismiss Bishop's intemperance and his kneejerk rhetoric that the "West is being abused," he was the new chair of the House subcommittee on public lands; he controlled which legislation made it to the House floor. Having helped ignite the uproar in the region over the possible increase in the number of protected wildlands, Bishop was highly unlikely to reverse course. He never let the Angeles and San Bernardino National Forest Protection Act out of committee.

However predictable, this was a blow to Wayne Steinmetz and the many others who love this jagged terrain's raw beauty.

Forget the Garden of Eden

Here's a problem with wilderness, or at least with our conception of it. We tend to think of it as unmanaged and, by definition, unmanageable—a place apart. There is the built environment, where we live, and the natural landscape, where we do not. The western origins of this dichotomy may date back to the Garden of Eden; its more contemporary sources can be traced to nineteenth-century European Romanticism and its American analog, Transcendentalism. Our Holy Trinity of Wildness—Emerson, Thoreau, and Muir—constructed a cultural defense of unspoiled beauty that has created a robust literature and poetics, been institutionalized in the national parks system, sparked the creation of wildlands organizations like the Sierra Club, justified the passage of and found sanction in the 1964 Wilderness Act, and launched a thousand glossy calendars glorifying the great outdoors.

Yet these social goods and the ideals they depend on come with an important qualification. If *Homo sapiens* has no place in nature—evicted with Adam and Eve—what is our responsibility to the places we profess to love so much? "Only people whose relation to the land was already alienated could hold up wilderness as a model for human life in nature," historian William Cronon has argued, "for the romantic ideology of wilderness leaves precisely nowhere for human beings actually to make their living from the land." This being so, our conception of wilderness "can offer no solution to the environmental . . . problems that confront us."

Yet how does this provocative argument work on the ground? I wondered about this while reading of the dormant status of Sycamore Canyon Park in Claremont, the historic college town on the eastern edge of Los Angeles County. Established in 1972, the 144-acre wedge of a park lies north of the Thompson Creek flood-control channel and is overlooked to the west by a lush, high-end development called Claraboya before it rises into the rough folds of Sunset Peak. Dotted with sycamore, oak, and eucalyptus and studded with chaparral, cacti, and manzanita, it is a foraging site for mule deer and coyote. Trails link its rolling terrain to the nearby city-owned, 1,600-acre Wilderness Park, and through it to the Angeles National Forest.

None of this environment meets the classic definition of wilderness, in which the human imprint is nonexistent. Ranching had already altered its ecosystem. Where cattle once grazed, cars roam the hardened streetscapes that come with suburbanization; much of the water that flows down Sycamore Canyon is immediately captured in a concrete ditch, an essential diversion if you want to construct residences on an alluvial floodplain. Through the high-tension power lines arcing overhead you can glimpse cell towers spiking along a distant ridgeline like so many porcupine quills.

Despite this durable evidence of our heavy hand, nature has not been exiled. Just ask anyone who had to flee the wind-whipped Padua Fire, an outbreak of the larger 2003 Grand Prix conflagration. It torched the brush-choked and wooded hills of northern Claremont, incinerating sixty-six homes; ashes swirled into the community like snow. Sycamore Canyon Park was among the spaces that burned with great intensity; it has been fenced off ever since. When a *Claremont Courier* reporter asked about this lengthy hiatus in spring 2011, a city official replied, "We needed at least six years to allow the site to regrow and reestablish itself."

That's the reflexive power of the idea of wilderness—only the wild can rehabilitate itself, only its growth is good growth. Nature knows best.

Unless you get sued. Then wilderness rhetoric can become something of a fig leaf. Hit with a series of post-fire lawsuits alleging that the city had failed to clear brush immediately adjacent to subdivisions that abutted its wilderness parks, in 2005 Claremont began to run goats in Sycamore Canyon to reduce its fuel loads. This biological control was supplemented with more mechanical interventions. In May 2008 a cash-strapped city council expressed relief that the Los Angeles County Fire Department would absorb the costs of cutting down and removing seventy eucalyptus trees killed in the fire.

Since then the city has intensified its managerial presence. In early January 2011 it announced that in collaboration with the Los Angeles Conservation Corps and other public agencies and nonprofit organizations, it would launch a yearlong, full-scale restoration of Sycamore Canyon Park. Nonnative vegetation will be removed, including 140 eucalyptus trees, and indigenous oaks and sycamores will be planted in their place. A new trail will be cut, switchbacking across a ridge to intersect with the dense network of paths and fire breaks that fan out across the southern face of the San Gabriel Mountains. Even the canyon's streambed may be altered, all with the goal of enhancing what city planners declare is the park's "unique and beautiful" environs.

The Sycamore project thus represents in small form the larger culture's uneasy compromise between our professed dedication to the preservation of wilderness and our unwillingness to leave well enough alone, between our rhetoric and action. It also continues the troubling wilderness-civilization dichotomy that has crippled the creation of an environmentalism that can lead us to embrace our obligations to repair what we have broken and our responsibility to integrate ourselves more carefully into the land that nurtures us.

Let It Be

The kangaroo rat is a misnomer. It is neither a marsupial (though it hops like one) nor a rat (though its facial features resemble one). Yet its inexact moniker is the least of the tiny, seed-eating rodent's worries. Unlike either animal whose name it bears, the kangaroo rat is endangered.

Its precarious situation is a consequence of the ecological niche it occupies, for the twenty-two species of the genus Dipodomys principally make their home in the deserts, arroyos, and washes of the Southwest. They are beautifully adapted to this harsh terrain. They have the ability to convert dry seeds into water; their kidneys are so efficient that they secrete little liquid; and unlike other animals, they do not need to pant or sweat to remain cool. By day these furry creatures live in burrows, avoiding the blistering heat; they venture out only as the night falls. Alas, that's when their predators are on the prowl, and they are not particularly adept at avoiding them—though the hopping helps; so does giving birth to three litters a year. But neither the kangaroo rat's agility nor its reproductive energy protects it from the bulldozer's blade.

Over the past twenty years earthmoving equipment

has scraped clean thousands of acres of pristine coastal sagebrush habitat in preparation for the construction of houses, big-box malls, and the highways that tie suburbs to shopping. These diesel-powered engines of development have been particularly busy in the Inland Empire of Southern California, Arizona's Salt River Valley (home to Phoenix), and the fast-growing eastern portions of San Diego County—terrain that is prime kangaroo rat habitat. Prior to 2008, federal protection of the endangered animal had saved some of its territory. That changed in the Bush Administration's final year, when the U.S. Fish and Wildlife Service (USFWS), bowing to executive branch pressure and construction industry lawsuits and lobbying, reduced the amount of "critical habitat" required to maintain the kangaroo rat's presence in its historic range. In San Bernardino and Riverside Counties, this decision led to a sharp reduction—from 33,291 to 7,790—in acreage set-asides established in 2002 for a local subspecies, the San Bernardino kangaroo rat. The steep decline sounded the death knell for the diminutive mammal, a fact that could not have been lost on builders, construction companies, and sand-and-gravel quarry operators.

The Center for Biological Diversity, a nonprofit organization committed to the protection of endangered species and threatened habitat, joined the San Bernardino Valley Audubon Society and Friends of the Northern San Jacinto Valley in 2009 to file suit in federal court in Riverside on behalf of the local subspecies. Their argument was blunt: the USFWS's scientific analysis and policy prescriptions were unsound and deeply flawed. U.S. District Judge Anne E. Thompson concurred, and in mid-January 2011, she struck down the 2008 USFWS decision, a judgment that ought to set a necessary precedent for the protection of other kangaroo rat species across the Southwest. "The Fish and Wildlife Service tried to gut critical habitat for the San Bernardino kangaroo rat," said Ileene Anderson, a biologist at the Center for Biological Divserity. "This latest court ruling gives this rare species a better chance at survival."

Notice she said a *better* chance at survival. Nothing is guaranteed, even with this important legal victory. As the economy recovers, developmental pressures will return, and they will focus, as they have in the recent past, on the alluvial fans that flow off the San Bernardino and San Gabriel Mountains. This rocky terrain is mined for materials critical to construction and is some of the last open space available for relatively easy conversion into commercial nodes, residential neighborhoods, and office parks. Drive along any of the freeways that crisscross the Inland Empire or along analogous landscapes throughout the arid region, and these building blocks of the modern American city flick by with monotonous regularity.

This sprawling development has not been good for the kangaroo rats and other flora and fauna whose lives are rooted in this desert scrub. To root them out is not good for us either. Start with our moral obligation to steward the planet, which demands a restraint we need to better cultivate. Surely that is one vital lesson in the aftershock of the 2007 economic crash. Not coincidentally it wiped out some of the malls and housing projects that had been slapped down on unspoiled desert-scapes after contractors and city officials promised they would bring unparalleled, enduring growth. This unsavory alliance and the speculative greed that underwrote its actions led to untold bankruptcies, innumerable foreclosures, and high unemployment. Growth for its own sake is not sustainable.

It is also fiscally irresponsible in another respect. Locating new communities and businesses in the middle of alluvial washes and floodplains is an expensive proposition. However cheap the housing appears to the individual buyer, the general public has to pick up the unstated but heavy costs associated with the construction of an interlocking network of dams, channels, ditches, and culverts designed to divert rampaging waters before they can crash into McMansions or McDonalds.

Keep these cautions in mind the next time you watch powerful winter storms spin into the coastal ranges or witness summer monsoonal rains hammer inland deserts. Preserving a generous expanse of kangaroo rat habitat might be the best and cheapest flood protection Southern Californians can buy.

Damaged Desert

Every time we drive a car, flick a switch, turn a faucet, go shopping, or grab a bite to eat, we are consuming nature. There is nothing shocking about this claim, but it is unsettling to see the consequences of our economic actions, to make the link between what we use and what we use up. To appreciate this ineluctable connection, head east into the Mojave Desert, where we are extracting the resources to drive Southern California's twenty-first-century economy, an industrialization of the desert that is predicated on making our lives easier but that will destroy this wild, beautiful, and arid land.

The Mojave's intense sunlight is being harnessed on mega-solar farms sprawling across vast stretches of this flat landscape. The region's geothermal potential—thanks to the deep rifts created by the Garlock, San Andreas, and other faults—will continue to be tapped. Its veins of rare earth minerals will be scooped out to build must-have hybrids and iPads.

The massive infrastructure needed to capture these sources of power and wealth, and the transmission and transportation corridors required to distribute them to consumers across Southern California and beyond, will speed up the degradation of this once pristine landmass that encompasses much of southeastern California and southern Nevada. Our hunger for all things material and the comfort we believe they will provide carries an unacknowledged price.

This stretch of high desert may be forced to bear another cost. When we arrive home from the mall with a haul and proceed to trash the sales receipts, packaging, and plastic bags, the accumulated refuse has to go somewhere. Until recently it has been jammed into canyons like the ones in the Chino or La Puente Hills, east of Los Angeles. These landfills are now nearly full, and more than twenty years ago Los Angeles County targeted Eagle Mountain in the Mojave between Indio and Blythe as its new dump.

Situated in a parrot's-beak of land that juts into the southeastern corner of Joshua Tree National Park, the town of Eagle Mountain is surrounded on three sides by protected wilderness. This curious setting and odd incursion has a decidedly political explanation. In 1936, when President Franklin D. Roosevelt created Joshua Tree National Monument (it gained park status as a result of the 1994 Desert Protection Act), Eagle Mountain was an integral part of it. Shortly after World War II, however, mining corporations successfully lobbied Congress, which pressured the Department of Interior to cut 265,000 mineral-rich acres from the national monument. One of the deal's key proponents and prime beneficiaries was steelmaker Henry Kaiser, who in 1948 bought the Eagle Mountain environs from the Southern Pacific Railroad, linked the mine to the regional railroad grid by building a fifty-one-mile spur line, and commenced digging iron out of the earth for processing at his Fontana foundry. When the mine played out in the early 1980s, the ghostly remains of his model company town and a traumatized terrain were left behind; the abandoned open-pit mine left a massive wound 4.5 miles long and 1.5 miles wide.

The huge scored chasm soon caught the attention of Los Angeles County planners who believed it would nicely answer the burning question of what to do with the metropolis's swelling volume of trash. Negotiations with Kaiser Ventures, LLD and Riverside County proceeded quickly. Federal regulatory agencies either sanctioned the project or their opposition was "duly noted," and proponents rolled up support by dangling the tantalizing job prospects (more than 1,300 would be employed) and economic growth (an estimated $3 billion would be pumped into regional coffers). The project seemed too good to be true.

It was. Eagle Mountain's proximity to Joshua Tree National Park quickly raised a host of red flags for environmental activists and park defenders. Pointing out the obvious—that landfills are notoriously hazardous—they worried about the deleterious impact an endless stream of garbage-packed trains and trucks would have on air quality and groundwater. Diesel-fueled compactors, which crushed 20,000 tons of trash daily, would further befoul the class 1 airshed and send aloft ballooning material to litter the park. The noise, stench, fumes, and high-intensity lighting would erode park visitors' wilderness experience and undercut the capacity of indigenous (and often endangered) flora and fauna to flourish. Seth Shteir of the National Parks Conservation Association declared that the landfill, an ecological disaster in the making, was "like putting a sewer next to the Sistine Chapel."

Protecting the sacred from the profane is the life's work of Donna and Larry Charpied, a pair of tenacious residents of the small Desert Center community. They moved to the Mojave in the 1980s, soon learned of the landfill project, and have been battling it ever since. Self-taught in the Byzantine-like intricacies of county, state, and federal law and increasingly savvy about the interlocking relationships between private powerbrokers and governmental regulators, the indefatigable duo has filed lawsuit after lawsuit to expose the Eagle Mountain project's unexamined environmental consequences. They also challenged a critical land-swap between Kaiser Ventures and the U.S. Bureau of Land Management deemed essential to the project's realization. Alone and in collaboration with the National Parks Conservation Association, the Riverside-based Center for Community Action and Environmental Justice, and the Desert Protection Society, the Charpieds have won every case.

Their most recent victory came in summer 2010, when the 9th Circuit Court of Appeals refused to overturn a lower court's rejection of the land swap. Although Kaiser Ventures would go forward with an appeal to the U.S. Supreme Court, its chances of success dimmed when the U.S. Department of the Interior decided not to support this last-gasp legal maneuver.

Yet even as Donna Charpied breathed a sigh of relief to local media—"Thank God it's over"—another controversy had surfaced. In early February 2011 the Federal Energy Regulatory Commission started taking public comment on a proposed hydroelectric plant that would turn the Kaiser pit into a two-level reservoir. By pumping millions of gallons out of a local aquifer and flushing them through a complex underground storage system, the facility would generate 1,300 megawatts of electricity, lighting up Southern California.

There is, apparently, no end to our appetite for the Mojave's resources, ensnared as they are in the web of our all-consuming desires.

Step Back

In winter 2011 Ventura County seemed to be channeling its inner King Canute, the legendary Danish conqueror of the British Isles. A millennium ago this powerful monarch is said to have commanded the waves to stop and the tides to cease. To the amazement of his fawning courtiers, Canute's words had no impact on the crashing waters, but his subsequent comment, however apocryphal or staged, has resounded across time: "Let all men know how empty and worthless is the power of kings."

The Gold Coast of Southern California seemed to have taken the humbled Dane's insight to heart. Ventura County ripped out a parking lot and concrete bike path that abutted Surfer's Point and moved these amenities further inland. The project, officially known as the Surfer's Point Managed Shoreline Retreat, is by name and action a deliberate acknowledgment that we too cannot control the sea.

This small act is part of a larger reconception of the human presence along the California coast, especially the heavily populated stretch from Santa Barbara to San Diego. Coastal engineers are no longer convinced that the best protection against the Pacific's tidal energy is to bring in the heavy armor. Where they once built seawalls, erected jetties, and laid down riprap, certain that these technologies would repulse waves' erosive power or redirect sand to preferred locales, they are now adopting a softer approach. Born of decades of research into the dynamic interaction between the ocean and the shore, and rivers' critical role in maintaining healthy beaches, scientists, planners, and enlightened government officials are beginning to enact innovative policies that will free nature to do its work.

Encompassing much of this thinking is the 2010 California Beach Erosion Assessment, whose central thrust is that Southern California beaches—which attract tens of millions of sun-worshipping visitors a year—are in deep trouble as a direct result of human actions that "have substantially altered the natural movement of sand and drastically reduced the natural supply of sediment to the coastline, thereby modifying seasonal beach building and erosion cycles." Upstream dams and debris basins, so valued for their capacity to control rampaging floodwaters, rob coastal areas of material nourishment. Every channelized river and stream, like the massive amount of impervious asphalt sprawling across the region and the incessant sand mining in local alluvial fans and riparian systems, has combined as well to "substantially decrease the supply of beach-compatible sediment provided to the coastline." Toss in the innumerable groins and breakwaters that litter the roughly 300 miles of regional coastline, and it's no wonder many beaches are starved for sand.

None of these interrelated issues are unique to California. The Eastern Seaboard and Gulf Coast suffer similarly, as do the European and Mediterranean, South African, Mexican, and Australian shores. Wherever urban development and intense beachfront recreation converge, you will find concrete infrastructure designed to enhance the tourist experience that is also responsible for compromising the shape-shifting terrain. In our rush to protect what we love, we have undercut these sandy objects of our desire.

This unsettling situation is further destabilized by climate change, which is elevating sea levels around the globe. The Arctic and Antarctic ice sheets are melting faster than any model has predicted. The sparkling beaches of the Cook Islands and other Pacific atolls have begun to slip beneath the rising waters. Facing inundation, too, are mega-cities such as Mumbai and Lagos. The injustice is staggering. These communities contributed precious little to the greenhouse gases that have heated our atmosphere but will pay a heavy price, part of which is recognizing that no amount of concrete can stabilize their imperiled ground.

That lesson should not be lost along the more secure Southern California coastline, and it's the reason Ventura County offers a small but important solution: pull back. Retreat from the beach, and as we withdraw tear out the fixed structures and hardened surfaces that have made our coastlines so rigid and vulnerable. We should also admit that cities built at or below sea level—think New Orleans or Galveston or Corpus Christi—are in jeopardy and devise ways to facilitate their rebirth on drier land. Doing so will require a political will and financial investment that beggars the imagination, but we don't have any other choice.

Maybe then we will be ready to admit that humanity cannot command or control nature, an arrogance King Canute decried long ago at the edge of the sea.

Net Loss

Come winter, the steelhead should be running up Southern California's many flat-bottomed rivers and creeks. They once took their cues from the powerful storms that roll ashore between November and March, dumping heavy rain on upcountry watersheds. When the mud-soaked rush swept downstream, breaching low-water sandbars in coastal estuaries before churning into the ocean, steelhead trout knew it was time to spawn.

In the past *Oncorhynchus mykiss irideus* took full advantage of this mix of meteorology, hydrology, and biology to nose into the South Coast's rivers, from the Santa Ynez River and Mission Creek to the Ventura and Santa Clara; from the Los Angeles, San Gabriel, and Santa Ana to the Santa Margarita, San Mateo, and San Luis Rey. Large-bodied (steelhead can pack more than fifty-five pounds on a forty-five-inch frame) and beautiful (their dark olive back is delineated from their iridescent silver underbelly by a luminous pink stripe), these powerful swimmers needed every ounce of storm-fed streamflow to help them navigate far up to the perennial, cool freshwater pools in coastal-range canyons.

Their difficult passage was a kind of homecoming, for after several years at sea the steelhead were returning to the sites of their birth. Like salmon, steelhead trout are anadromous; born in freshwater, within twelve months or so they experience a physiological transformation known as smoltification that allows them to survive in the ocean's harsher chemistry. The same winter floods that will guide them back to their high-country spawning grounds propel their migration toward saltwater.

This cycle endured for thousands of years, a piscine movement between riparian and marine habitats that marked each ecosystem's enduring health. Healthy they were, too, if one can judge by the teeming numbers of steelhead that used to ply Southern California waterways. The Santa Ynez in Santa Barbara County, for example, is thought to have supported the largest run of steelhead south of San Francisco. In winter 1891 observers stationed at its mouth tallied "great schools of young salmon" heading upriver.

It has been a long time since anyone witnessed its like. Starting in the early twentieth century, more and more people have crowded into the Southland's valleys, foothills, and highlands. To ensure that these varied terrains serve our needs, we have manipulated the environment to such a degree that we have wiped out the indigenous steelhead population.

Our weapon of choice was concrete. Those pounding rains, whose annual runoff flipped on the steelhead's reproductive switch, were a different kind of harbinger for the human settlements sited along these same watercourses. To protect against damaging floods, federal, state, and local agencies funded the construction of dams and an interlocking network of channels, ditches, and drains to impound and divert stormwater. In 1920 the Santa Ynez's Gibraltar Dam sealed off a portion of that river's historic spawning grounds and the 1953 Bradbury Dam delivered the coup de grâce; as fatal was the 1948 Mitilija Dam sited on the Ventura. These and other dams impeded and confused the steelhead's migratory impulse.

Surely their confusion was magnified by extensive groundwater pumping for thirsty farms and urbanites that reduced streamflow, bridge-and-highway infrastructure that ripped apart riparian ecosystems, and bulldozed sprawl that buried seeps, wallows, and springs.

No less befuddling is the human ambition to rearrange these rivers in our image. We worship the linear and find comfort in a geometric order that these alluvial streams defy. Following the path of least resistance, their course through the gravelly soils has varied depending on the season. After the Los Angeles River squeezes through the Glendale Narrows, for example, its path to the sea could shift north or south across a vast floodplain; its mouth has followed suit, ranging from Santa Monica Bay to the San Pedro. At one point it even flowed into what is now Long Beach harbor. Steelhead trout had no trouble adapting to this hydraulic variability, but real estate developers, engineers, and urban planners have not been so flexible. Their sense of propriety and property, bound up with their abiding faith in technology's ability to box up the wild, led them to pour yards upon yards of concrete to turn the fifty-mile river into a culvert straight and true.

This is not a landscape steelhead could love. By the mid-twentieth century its southern populations had begun to crash; sixty years later the U.S. Fish and Wildlife Service estimates that its numbers have declined precipitously, by over 99 percent. What worked for us has killed them.

This deadly outcome is symptomatic of a larger conundrum. To restore historic steelhead runs will require reconstructing the historic conditions that nurtured their anadromic life cycle. That's not going to happen, in large part because few of the dams—and the reservoirs they created—will be removed. But this reality has not stopped people from trying to revive the world the steelhead once knew. Buoyed by the passage of state and federal clean water legislation, encouraged by laws protecting endangered species and wild and scenic rivers, and optimistic that ecological analyses can properly guide restoration projects, grassroots organizations have been pushing, suing, and working with governmental agencies to repair what earlier generations tore asunder.

Since the early 1990s the Environmental Defense Center (EDC) of Santa Barbara has fought to bring steelhead back to local waters. Most recently, in collaboration with the Army Corps of Engineers and city planners, it has helped redesign the concrete channel that frames Mission Creek, essentially punching holes in it to slow its flow. Plans call for the creation of step pools and rock weirs to facilitate fish migration. In Ventura County the Matilija Coalition has worked diligently to remove the eponymous dam blocking the Venture River, modify bridge structures, and regenerate the riverside canopy of oak, sycamore, and willow. Similar efforts are under way at the mouth of Malibu's Solstice Canyon Creek. EDC's Brian Trautwein spoke to the hopes of many Southern California activists when he predicted that on Mission Creek "we'll restore passage for migrating steelhead and bring this remarkable and resilient species back from the brink of extinction."

Less sanguine news has surfaced in northern San Diego County. Beginning in 2003 the California Coastal Conservancy and its partners have spent hundreds of thousands of dollars trying to revive the San Mateo Creek steelhead population. Despite killing tens of thousands of invasive fish and frogs, in late January 2011 the conservancy threw in the towel. Its failure is attributable to a number of site-specific problems. Camp Pendleton blocked its efforts to restore the creek's lower stretch, a critical hindrance made more difficult by the group's inability to extirpate steelhead predators that continued to stream into the creek's upper watershed from private ponds. The irrigation district carped that this particular setback revealed a systemic flaw in the idea of restoration. "If on a smaller, more manageable watershed like the San Mateo they are unable to make progress," Don Smith, director of water resources for the Vista Irrigation District, told the *San Diego Union Times*, "what makes them think they are able to make progress on a much larger and more complicated watershed?" The conservancy's response is clear; it has shifted its funds to other sites because it remains convinced it can return the steelhead to its native habitats.

So are the Friends of the Los Angeles River. Under the inspired leadership of poet Lewis MacAdams, its members have worked tirelessly to help Angelenos reimagine this massive concrete channel as a natural river. To achieve this first required public access to the river's steep and hardened banks, locked behind chain-link. To cut through the fence and open the space, the friends' group and its allies envisioned a series of pocket parks tied together by walk-and-bike trails. One of these, set within the Glendale Narrows and now planted with indigenous flowers, grasses, and trees, is Steelhead Park. Its naming is not by happenstance. "It took more than forty years to screw the river up," MacAdams wrote in the *Whole Earth Review* in 1995. "I'm sure it will take more than forty years to bring it back to life again. From the beginning we said that not until . . . the steelhead trout were swimming up the river to spawn . . . would Friends of the Los Angeles River's work be done."

Until then this tranquil spot must stand as mute testimony to the thrashing, silvery surge that once roiled Southern California's free-flowing rivers.

Shady Dealings

Every March California celebrates Arbor Day, a national campaign to spruce up the country by releafing its streets, adding shade overhead, and injecting a little arboreal calm into our busy lives. So it is probably not the best time to take a poke at Los Angeles Mayor Antonio Villaraigosa's Million Tree Initiative. Actually, there's not a better moment to examine its unchecked presumption that trees will save us. Can this, should this dry place support all that projected growth?

Launched in 2006 with considerable éclat, the program got a quick rhetorical jump-start. "I am committed to making Los Angeles the largest, cleanest, and greenest city in the United States," the mayor declared shortly after his first election. "Today only 18.09 percent of Los Angeles is covered with trees. I ask all Angelenos to work with me and engage in this effort to grow our canopy cover for the future by planting one million trees today."

Villaraigosa's ambition was heartfelt, his gesture full of symbolic import. By this policy he expected to make a definitive statement about his environmental convictions and the priority they would hold during his tenure. His bold aspiration even had some eye-popping scientific data to support it. Urban forest researchers at the U.S. Forest Service suggested in 2007 that planting one million trees would reduce stormwater runoff, decrease the city's carbon footprint, cut the use of air conditioning (and thus electricity), and make for a more beautiful community, with all the attendant psychological and health benefits that would flow from an enhanced and expanded tree canopy. Over thirty-five years, the per tree savings would range from $1,100 to $1,600, an estimate that totaled a staggering $1.1 to $1.6 billion. What politician could turn his or her back on such a return rate?

Perhaps this one should have. Or at least Villaraigosa might have paid a bit more attention to the concerns local tree activists raised about the program's feasibility, such as the worry that it would prove impossible to calculate the real—as opposed to the potential—payback for planting such an astounding number of trees. How, some asked, would the city determine if the many seedlings it freely handed out at parks, educational fairs, and schools made it into the ground? And if they were planted, how would anyone know whether they survived? The whole scheme resembled a shell game.

Another gamble involved the water needed to nourish so many trees to maturity. Nowhere in the Forest Service's calculations or the city's analyses was there mention of this potential complication, which is a bit surprising in a city with an annual rainfall of less than fifteen inches, a city that annually imports tens of thousands of acre-feet of water to sustain itself and whose drawdown of Northern California streamflow is constantly threatened due to endangered species concerns in the Sacramento–San Joaquin Delta. To green up the city would require a good deal more white gold, making Los Angeles even more complicit in statewide environmental degradation.

There was one aspect of the Villaraigosa campaign that seemed beyond reproach. As proposed, it would correct an environmental injustice reflected in the distribution of trees across Los Angeles. The Forest Service's analysis of the city's tree canopy revealed that such wealthy enclaves as Bel Air, Beverly Hills, and Hollywood Hills have a thicket of woods that shade their swimming pools and cool off their movie stars. But head south down Robertson, La Cienega, or Fairfax, and even before you roll beneath the thunderous Santa Monica Freeway the trees thin out and the sun's heat intensifies. As the federal researchers noted, the percentage of canopy (what they dub TCC), when set within city council districts (CDs), "varied from lows of 7 to 9 percent in CDs 9 and 15 (Perry and Hahn) to a high of 37 percent in CD 5 (Weiss)." This led them to conclude that canopy "was strongly related to land use. As expected, low-density residential land uses had the highest TCC citywide (31 percent), while industrial and commercial land uses had the lowest TCC (3–6 percent)." The same troubling results emerged when comparing neighborhood councils. The canopy density was vastly different between Bel Air–Beverly Crest (53 percent), Arroyo Seco (46 percent), and Studio City (42 percent) and those neighborhood councils with the lowest TCC, including Downtown Los Angeles (3 percent), Wilmington (5 percent), and Historic Cultural and Macarthur (6 percent). The poorer the community or the more industrialized the landscape, the fewer the trees.

Righting that wrong by greening the city's most shade-starved and concrete-hardened districts makes good sense and good politics. Yet this assumes that the wide-scale planting of trees throughout Los Angeles is ecologically sound and appropriately sustainable, a query particularly appropriate on Arbor Day, because for more than 120 years the Arbor Day Foundation has promoted tree planting—everywhere and anywhere—as an unalloyed social good and environmental benefit.

The upshot, though, has been a homogeneous landscape. Because trees are beneficial, activists have asserted, they should be given root even in terrain that does not naturally support them. It is telling that Arbor Day got its start in largely treeless Nebraska; telling, too, that its guiding genius, J. Sterling Morton, grew up in the humid East. Morton's tireless advocacy was designed in part to turn the prairie into a well-wooded land. Call it agro-imperialism.

Anglo-Americans unfurled this same imperial banner when they poured into late-nineteenth-century Los Angeles. Its coastal sagebrush ecosystems, so spare and sparse, seemed alien to those who had come of age in the

Midwest pineries, lush southern forests, or New England hardwoods. Because there is no place like home, they immediately set about domesticating the local environs. They dug countless holes in the rocky soil, inserted and patted down thousands of seeds and seedlings from their natal homes, and watered the young growth with abandon.

As they planted these trees, ornamental in name and function, in every town that sprouted up along the rail and trolley lines crisscrossing the valley floors, these migrants were Americanizing Southern California. Each tree became a symbol of the powerful civilization that had forcibly supplanted older communities, whether native, Spanish, or Mexican, which managed this landscape with a lighter hand. Environmental transformation thus was linked to political domination. Whites liked green streets; shade marked the arrival of the American Colossus.

This troubling history colors early twenty-first-century programs like the Million Tree initiative. True, its ideological arguments are framed in contemporary scientific discourse—for example, that these many trees will sequester so many tons of carbon in the fight against climate change. Still, the idea that we must increase the region's canopy cover flies in the face of more compelling scientific constraints—the levels of heat and precipitation, quality of soils, and amount of light—that determine a tree's capacity to grow. By ignoring these technical details, Villaraigosa's program evokes an earlier generation's deliberate rejection of environmental realities in favor of imported cultural norms.

Perhaps this willful disregard is inevitable in a city whose official tree is the coast coral, an introduced species from southeastern Africa. Really, we should know better.

Breathe Deep

The postcard on my desk is almost forty years old. Angelenos of a certain age will recognize it—a wide-angled aerial shot of the downtown core of Los Angeles and its much more modest skyline. Framed by the intersection of the Santa Monica and Harbor Freeways, the scene is muffled in brown smog. Barely visible in the deep background, poking above the thick toxic stew, is a snow-capped Mount Baldy, at 10,064 feet the tallest of the San Gabriels. "Greetings from Los Angeles," reads the arch caption.

I spotted the card in fall 1972, when I came to Southern California to attend Pitzer College, and immediately sent a steady stream of them to family and friends back East. They got its black humor, which I reinforced when I confessed (perhaps bragged) that although my dorm room was within five miles of Mount Baldy I almost never saw its bold outline.

Now I see it every day, often with stunning clarity, as if the entire range were etched out of a blue true dream of sky. How strange, then, that Republicans in Congress are maneuvering to gut the Clean Air Act, stop the Environmental Protection Agency from regulating greenhouse gases, and, in a special affront to Los Angeles, roll back the federal agency's ability to monitor tailpipe emissions. It's enough to make you gasp for air.

This regressive political agenda, designed to savage public health, ought to infuriate anyone who suffered through the dark-sky years that hung over Southern California like a pall. It took decades of fierce struggle on the local, state, and national levels to build the political capital and legislative clout to write the binding regulations, a battle that began in the late 1940s and is richly chronicled in Chip Jacobs's and William J. Kelly's 2008 book, *Smogtown*.

It took just as long to create and fund the federal Environmental Protection Agency (1970) and the local South Coast Air Quality Management District (1976). Neither organization had an easy birth. President Nixon reluctantly created the EPA under considerable pressure, and Gov. Ronald Reagan twice vetoed the creation of the South Coast district, which only came into being with a stroke of Gov. Jerry Brown's pen.

We have blue skies—when we have them—only because of the robust regulatory regime that emerged from this fraught politics of smog. Nothing else accounts for the steady uptick in Southern California's air quality. What my vintage postcard, in its didactic back text, asserts were the central contributing factors to the region's then poisonous air remains true: "Millions of people driving millions of cars plus temperature inversion provide Los Angeles with a near perfect environment for the production and containment of photochemical smog." One result of this disturbing mix of technology and meteorology, it warns, is that the "LA Basin inversion acts as a giant lid over the smog, inhabitants, and visitors."

You know what happened next. As the heated air compressed, it produced intense stomach-souring, eye-burning smog that made lives miserable in the San Fernando and San Gabriel Valleys, choked downtown, and obscured the Hollywood sign. Much of this pollution ended up in the Inland Empire, pressed east by the prevailing on-shore breezes, where it mixed with what my friends and I dubbed the "smell of the ick," a nauseating brew that Kaiser's Fontana steelworks routinely vented. On such days parents kept their kids out of school, athletes trained indoors, citrus growers and sugar beet producers watched in dismay as their crops withered, and the elderly and young crowded into doctors' offices and hospital emergency rooms with throbbing heads and shortness of breath. Los Angeles was flatlining.

The citizenry resuscitated it. They were mobilized, as Jacobs and Kelly recount, by groups like the League of Women Voters and an ad hoc battalion of moms, poor and well off, who battled the city's tone-deaf power elite. Farmers and fathers added their voices to the growing outcry that *Los Angeles Times* cartoonist Frank Interlandi stoked in panel after panel. In one of these, a brutal 1967 creation, swirling black clouds drop a set of gagging Angelenos to the sidewalk. They collapse before a newspaper stand, its headline screaming, "Smog Control Board to Act; Deaths Spur Activity." Interlandi's contempt for ineffectual pols and their fatal indifference was palpable.

Also effective were the pathbreaking analyses of scientists at the Riverside Citrus Experiment Station (progenitor of the University of California, Riverside), who identified auto and truck emissions as the source of agricultural die-off. So, too, was the innovative work of Arie Haagen-Smit, a renegade Cal Tech biochemist who developed the first techniques to analyze smog's chemical composition and target its origins. His critical research ran afoul of the auto industry but gave protestors vital data to support their case in the streets, at city hall, and before the judiciary.

The combination of grassroots organizing, scientific investigation, and a galvanized public proved potent. However slow the process may have been, however incomplete its results, without this engaged anger the quality of life in Los Angeles—and a lot of other American cities—would have been greatly diminished. A recent retrospective EPA study charting the savings in lives and money that can be credited to the 1970 and 1977 Clean Air Acts shows that without these two bills, by 1990 "an additional 205,000 Americans would have died prematurely and millions more would have suffered illnesses ranging from mild respiratory symptoms to heart disease, chronic bronchitis, asthma attacks, and other severe respiratory problems."

Of special interest to Los Angeles and other auto-centric communities is the EPA's finding that "the lack of Clean Air Act controls on the use of leaded gasoline would have resulted in major increases in child IQ loss and adult hypertension, heart disease, and stroke." The regulations' total fiscal benefit ranges from $6 trillion and $50 trillion, making it mind-blowingly obvious how price-less these legislative initiatives have been.

And how utterly foolish their Republican detractors were then—and are now. Indeed, that's one reason I have kept the iconic smog postcard, taping it to every desk I've ever worked at. Once a sick joke, a reminder that back in the day we had neither the political desire nor the survival instinct to protect ourselves from ourselves, the card has morphed in meaning. Now it also evokes how far we have come, how indebted we are to the men and women who fought against the big car companies, rebutted their kept scientists, and took down a legion of political lackeys. They convinced Congress that the environment mattered, that federal regulation and well-funded enforcement mechanisms were essential tools in the defense of the people's health, safety, and welfare.

We are the lucky beneficiaries of that generation's principled activism. It would be nice if our children and grandchildren could say the same about ours.

Pumped Dry

Wash your hands, irrigate crops, or cool an industrial process. If you do this anywhere in the Southwest, chances are the water comes from someplace else. To move it from where it has fallen as rain or snow to our taps, irrigation pipes, or conduits, state and federal governments have spent billions of dollars on a massive array of distribution systems, including the Central Utah and Central Arizona projects and the California Aqueduct. Despite their brilliant engineering, their remarkable capacity to push vast quantities of this captured flow over mountains and across deserts, these interlocking structures—dams and reservoirs, channels, tunnels, and pumps—contain a fundamental flaw. They were built in the mid-twentieth century when the region's population was considerably smaller and precipitation was more abundant.

That's no longer the case. Which is why before the 2011 celebration of World Water Day, the U.S. Center of the Stockholm Environmental Institute (SEI) released an ominous report on the water crisis confronting California and neighboring Nevada, Utah, Arizona, and New Mexico. Backed by arresting data, "The Last Drop" indicates that this region, home to some of the nation's largest and fastest-growing cities and the site of major agricultural production, is in deep trouble.

Yes, winter 2010 was cool and wet. The reservoirs were full, the rivers were running fast, and the upper elevations of the San Gabriel and San Bernardino Mountains, like the Sierra, Wasatch, and Rockies, were thick with snow. The apparent bounty was such that California Sen. Dianne Feinstein issued a press release demanding a greater allocation of water to some of her favored constituents (and donors)—the major agricultural interests in the Central Valley. "The disconnect in federal water allocations is the worst I've seen in years," she asserted. "South-of-Delta farmers are getting only 55 percent of contractual amounts, a shocking number when the state snowpack is as high as 165 percent. That is simply unacceptable."

Actually, what was unacceptable was the senator's presumption that state—and by extension, regional—water policy should be enacted during flush times, framed around short-term analyses, and devoted to a single interest. This kind of narrow thinking has set up Californians for the coming water-pinching years. Yet it almost seemed rational compared to some of the head-scratching claims of Republicans representing California in the U.S. House of Representatives. Consider Water and Power Subcommittee chair Tom McClintock, a former Southern Californian who was representing a district that runs from Sacramento to the Oregon and Nevada borders. His response to a season so rainy that the water-project pumps supplying Central Valley farmers were turned off due to lack of demand was to hold hearings designed to preach the virtues of "a policy of abundance," as if abundance is a quality the federal government can conjure up. Or Congressman Devin Nunes of Tulare County, who, when Sen. Feinstein and California's Secretary of Natural Resources John Laird opposed his legislation to undermine ongoing water policy negotiations, angrily tweeted that they had "move[d] quickly to starve valley of water."

That year's wet weather and the bizarre politics it inspired did not change the fact that the long-term news is grim. According to the SEI, the five southwestern states were already engaged in unsustainable practices. Individually and collectively they were withdrawing more water than nature was replenishing, overpumping ground- and surface water supplies to meet the rapidly increased demand. The Central Valley is a case in point. Its groundwater supplies have been disappearing much more rapidly than predicted. Between 2006 and 2010 farmers pumped more than 12.8 million gallons of water, which works out to 40 million acre-feet or, as University of California (Irvine) hydrologist Jay Famiglietti advised McClintock's subcommittee in spring 2011, enough to bury half the state under a foot of water. Years of steady rain could not begin to make up the loss.

Clearly the pressures on regional water systems will only intensify if, as predicted, agricultural production of water-intensive crops remains the norm across the Southwest, and if Las Vegas, Los Angeles, San Diego, Phoenix, Tucson, Salt Lake City, San Antonio, and Albuquerque grow as quickly as they have over the past twenty years. According to one of SEI's projections, which assumes that population and income growth will continue and holds constant current climate conditions, the overdraft is an estimated 260 million acre-feet. That's an unnerving number, especially when you consider that a single acre-foot is roughly the amount of water needed to supply two average American families for one year.

It becomes more worrisome when you factor in a modest change in climate that projects diminished precipitation between 2010 and 2110. Then the water shortfall amounts to a staggering 2,096 million acre-feet. Should current trends in global greenhouse emissions continue unabated—which the SEI report considers likely—the shortfall spikes to 2,253 million-acre feet.

This is bad news for Arizona, which draws 90 percent of its water from the Colorado River and local aquifers. It might be worse in Nevada, which has the highest per capita water use in the country. But the good citizens of the Golden State are really up against it. To survive the next century at present rates of extraction, California will need *three times* the currently available groundwater. In a state already saturated with dams, no amount of reservoir construction will meet that need (one reason opponents were able to quash former Gov. Arnold Schwarzenegger's

2007 proposal to spend $4 billion to build two new dams). In a rational universe this data would silence the squabbling rhetoric of those who in 2011 were trying to grab additional supplies without thought to the drying future.

Saner is the conviction of the Oakland-based Pacific Institute that the state can achieve a sustainable balance between water supplies and demand. First we have to admit that the next century can look nothing like the last. "Unconstrained and unmanaged growth in southwestern cities and suburbs can no longer be accommodated from the perspective of water supply," noted the institute's president, Peter Gleick, in an article in the Proceedings of the National Academy of the Sciences. Neither can agriculture continue to operate as if nothing has changed. "The irrigation of certain crops in certain places no longer makes sense, even with economic subsidy." As for urban consumers, they "must be more attuned to the hydrologic realities of the region."

The impetus driving these critical reconsiderations is climate change. Adapting to its impact on a strained water supply, Gleick and his colleagues argue, requires the implementation of high-efficiency measures. The goal is to reduce consumption by irrigation-heavy agriculture and thirsty cities through technological and fiscal incentives. After all, the Southwest—urban and rural—was built on cheap, subsidized water. Many California farms receive enormous quantities today simply because the law acknowledges the old principles of "we were here first" or "we hired the right lobbyists." That approach is already unsustainable in the face of a growing population and a decreasing water supply.

Some first steps have already occurred. Per capita water use in Los Angeles is roughly the same today as it was forty years ago, a consequence of elevated pricing and rigorous conservation. The same is true across the nation. "Total water use in the United States," Gleick observes, "was less in 2000 than it was in 1975, yet population and gross domestic product over that same time increased." We've done more with less.

SEI researchers are doubtful that these efforts amount to much. Without radical decreases in use, particularly in the agricultural sector, which uses 78 percent of the region's developed water but produces less than 2 percent of its gross domestic product, life here will be very tough.

If Californians want to continue to call this rough, dry, beautiful land home, if they want the pumps to bring up water, not dust, then they are going to have to adapt—now.

Course Correction

To thumb through Fred Eaton's faded scrapbook, tucked away in special collections at the Claremont College's Honnold/Mudd Library, is to recall the incredible ambition that drove Los Angeles to wrest water out of the Owens Valley and channel it more than 200 miles south. This bounteous flow would help power the city's growth for much of the subsequent century while relegating the eastern California high desert to a backwater.

Eaton, who had served as city engineer and mayor, led this November 1905 fact-finding mission, which included city council representatives, civic leaders, and journalists. For the past several years agents acting on their own and for the city had been buying up land and the water rights they contained, and Eaton was among them. So he was in the sketchy position of cheerleading a process he was a prime beneficiary of.

No surprise that the photographic evidence of this tour and the daily diary accompanying it are almost entirely framed around the creeks, rivers, and springs the group encountered. An image of a very full Stevens Ditch is accompanied with text noting that it "carried 600 inches one half now owned by the city." ("Inch" here refers to the archaic term—a miner's inch—which was used to measure streamflow; a single inch meant the discharge was an estimated nine gallons a minute.) The next day the group visited the 50,000-acre Rickey Ranch, which Eaton had recently sold to the city and which was nearly "20 miles long, reaching across the entire valley." Among its resources was the prodigious Black Rock Springs (photographed in full flood), which produced an estimated 1,000 inches. The property, Eaton observed, "practically controls the entire water situation of the valley."

The interplay between the scrapbook's pictures and words reinforces the idea of capture, which these boosters underscored in their nightly congratulatory speeches. Eaton lauded "the enviable position in which the city of Los Angeles was so fortunately placed in having control of such a magnificent water supply."

Touting this gospel of wet wealth became their self-appointed task when they returned to Southern California, for they concluded that only the Owens Valley could provide the large water volume they believed necessary to slake Los Angeles's deepening thirst. Its growth took precedence to that of the towns they visited—Lone Pine, Independence, and Big Pine. Its boom depended on their bust.

In time Los Angeles would also grab control of a goodly portion of the Colorado River's icy flow and snatch a major share of precipitation falling on the northern Sierras. To move these disparate waters into Southern

California, the city and, later, the state spent billions to construct conduits, pumps, and aqueducts. The name of the local agency charged with developing these resources, the Department of Water and Power, could not have been more exact.

Yet this agency's clout, long tied to the operating presumption that the water of other peoples and places is there for the taking, and confirmed in the city's ownership of nearly 4 percent of Inyo County (home to the Owens River watershed), is not what it was. Over the years a series of state agreements and federal compacts has circumscribed the freewheeling exploitation that once defined Los Angeles's water policy, an imperial mien that nonetheless remains embedded in popular culture. Rural communities throughout the Great Basin, for example, buck themselves up as they struggle to stop an ambitious Las Vegas from siphoning their groundwater supplies by vowing to "Remember Owens Valley."

These grassroots activists should take note of recent federal court findings that serve as an additional breakpoint in Los Angeles's water history. Since 2007 Judge Oliver W. Wanger has issued a string of decisions about the deleterious impact that pumping water from the north to the south—and for Central Valley irrigators—has had on the Delta smelt. Although he has been inconsistent in his support of the Endangered Species Act, oddly asserting that Congress never intended to "elevate species protection over the health and safety of humans," and although in December 2010 he threw out portions of the biological opinion guiding calculations the U.S. Fish and Wildlife Service used to restrict pumping, Wanger has not (yet) reversed the larger logic: the dwindling number of smelt has and will trigger limits on the amount of water flowing to Southern California cities and rural agribusiness.

His decisions, even if equivocal and disjointed, have already impelled some communities and agencies to plan for a time when what they have taken as normal—a massive stream of water pouring into the region from elsewhere—will no longer be the norm. Santa Monica, for one, is rapidly pursuing water independence. In February, with the opening of a $60 million water treatment plant that treats polluted water pumped out of local well-fields, it announced that it could supply eight million gallons a day, approximately two-thirds of the city's daily consumption. "Given the risks and uncertainty of the California water supply," Mayor Richard Bloom said, "the only reasonable response for Santa Monica is to reduce its use of an imported water supply."

Others have come to the same conclusion. With its Alamitos Barrier Recycled Water Project, for example, the Water Replenishment District of Southern California, a regional groundwater management agency, will "replace the barrier's imported potable water demand with a recycled water supply, thereby improving water supply sustainability within the Southern California region."

Cutting imports from 64 percent to 20 percent is a sign that the agency is succeeding in its campaign for Water Independence Now. That there is such a strategy, complete with a snappy acronym (WIN), speaks to the growing cachet of similar projects across the region. Long Beach, the San Gabriel Valley Water Replenishment Project, and Orange and San Diego Counties are among those entities spending considerable money to shore up, recharge, and/or protect groundwater systems. By these innovative efforts, Southern California is reclaiming local sources of potable water after a century of neglect.

Though these reclamation projects are precedent setting, they are also echoes of a stay-at-home choice available long ago when Fred Eaton and the power elite schemed to get ownership of the distant Sierra watershed. Consider the actions residents of Pomona, LaVerne, Claremont, and Upland took to defend their rights against Los Angeles. In 1909 they formed the Pomona Valley Protective Association, charging it to manage and conserve local groundwater. Its mission has been fulfilled through its "spreading grounds." The association controls more than 1,000 acres of the alluvial fan that flows off Mount Baldy and through San Antonio canyon to the valley below. A system of ponds, dikes, and channels flushes water across the rocky terrain so it can percolate into the Canyon and Upper Claremont Heights Basins. Although the member communities are no longer water-independent, their century-long commitment to sustaining these aquifers puts them in an unusual position to achieve greater independence in the future.

This historic approach to water resource management represents the road early-twentieth-century Los Angeles refused to take. The plight of the Delta smelt makes it clear this is the path we must return to.

Mud Fight

Three hundred to four hundred dump trucks a day. Five days a week for three years. Assuming a fifty-week year, that amounts to 75,000 truckloads. Why are all those diesel-powered vehicles needed for so long? That's the amount of machinery and labor, according to a recent Los Angeles County Department of Public Works (DPW) report, required to haul away the 1.5 million cubic yards of debris that have built up behind Devil's Gate Dam, north of downtown Pasadena.

Much of this astonishing load is a result of post–Station Fire flooding in winter 2010, when an estimated 1.5 million cubic yards caromed downhill into the sealed-off canyon. Now is the time to dig it out, DPW asserted, an assertion county supervisors did not accept.

Neither did some nearby residents, who protested the noxious fumes and swirling dust that would be kicked up by the massive earthmoving equipment and an endless stream of trucks; they were equally opposed to the destruction of wildlife habitat and recreational space that have built up on that gravelly spoil. Others challenged DPW's right to act without assessing the short- and long-term consequences of its actions. As Tim Brick, who directs the Arroyo Seco Foundation, pointed out to the *Los Angeles Times*, "Everybody else has to do environmental impact reports, why shouldn't the county?"

He has a point. So does the *Times*'s editorial rebuttal, which called out the supervisors: "They have received clear warning from the experts in this matter that the current interim measures won't do the job if there is catastrophic runoff. Delay is dangerous, and protecting the communities downstream from the dam must take priority."

Fair enough. The fact of the dam's presence and its centrality to modern life—indeed to humans' ability to live in this region—cannot be gainsaid. Something has to be done. The central question, though, is not what must be done and when, but on what basis. I'm not trying to split hairs. It seems essential, epistemically and ethically, to recognize that the argument DPW and the *Times* have deployed depends on a couple of unexamined suppositions.

Begin with that tantalizing "if" in the *Times* editorial, as in "if there is catastrophic runoff." Did the newspaper hope its readers would miss the huge qualification it hung on that word? Did it, and the department it is flacking for, expect readers to forget that these two institutions have always pursued this strategy to manipulate flood-control policy and politics? The agency that constructs and manages debris basins and the communal cheerleader of all things constructed—they routinely insist that nature can and must be defied, that human ingenuity and technology can harness the San Gabriel Mountains' millennia-old processes for flushing billions of cubic yards of rocky debris into the valley below, forming alluvial fans that are miles wide and over 900 feet deep. We can move mountains.

This same ambition explains the existence of Devil's Gate Dam, near the 210 Freeway and Oak Grove Drive in Pasadena. Built in 1920, it was designed to defend Pasadena and other communities from thunderous flows of mud, boulders, rocks, and plant material that periodically sluice down the Arroyo Seco. Selecting the structure's site was determined in part by the infamous 1914 flood.

Late that February, after a furious storm, a massive debris slug surged down the high-country watershed and roared through the canyon's narrow "gate" before slamming into Pasadena. It inundated flanking neighborhoods; gouged out chunks of streetscape; carried away homes, bridges, and buildings; and killed forty-three people.

At that moment Pasadena and the county had an opportunity to ask themselves how best to live in this place. They might have drawn on the experience of the Tongva and Spanish peoples, who deliberately did not occupy the flood zone, and pulled back. They could have shrugged, accepting that deadly debris flows were part of the price for inhabiting the Arroyo Seco watershed. Instead they decided to try to minimize those costs by erecting the Devil's Gate Dam. A subsequent flood in 1916 sealed this devil's bargain. Pasadena granted the county an easement to build a dam at the gate, and a year later voters passed a controversial bond measure to fund the project; the dam was completed in three years for $483,000 (about $5.3 million in today's dollars).

The dam did its job, and the fact of its success encouraged Pasadena, like a lot of other Southern California towns that in time would clamor for dams, basins, and cribs, to push development into once unbuildable floodways. The iconic Rose Bowl (1922), set within the Arroyo Seco, could not have been conceived without Devil's Gate. And the DPW and the *Times*, with unintended irony, have used the legendary gridiron's precarious position to press their case for the immediate clearing of debris from behind the dam. Within ten minutes of a calamitous flood topping the dam, we are assured, the stadium would be awash.

Other post-1920 infrastructure was also identified in DPW warnings about the consequences of not acting immediately: "Many areas near the Arroyo Seco, including homes, horse stables, and the 110 Freeway in northeast Los Angeles, South Pasadena, and Pasadena are at risk of flooding."

Nothing about this problematic situation is unique to Los Angeles. The earthen and concrete levee systems that line the Mississippi River have allowed developers to build commercial, industrial, and residential nodes in the low-lying sections of Memphis, Vicksburg, and New

Orleans—all of which, if only briefly, went underwater in the Great Flood of 2011. At roughly the same time Devil's Gate was constructed, imperiled Gulf Coast cities began throwing up barriers to hurricane-driven storm surges, and then they built right up against these hardened structures on the cheerful assumption that they were out of harm's way. The same thing happened in flood-prone Phoenix and San Antonio; their respective dams—the Roosevelt and the Olmos—encouraged the urbanization of once dangerous flood basins.

The result was predictably calamitous. Not all of these structures could withstand nature's fury at every moment, so each community has felt compelled to add redundant lines of defense to protect its original investment. They thickened and raised levees, boosted the size and strength of seawalls, and reconfigured and extended the grid of dams, channels, and diversion ditches. Yet many of these fixes accelerated the environmental damage and increased costs in an endless cycle. Most notorious is below-sea-level New Orleans, which absorbed billions of dollars before Katrina to shield it from disaster and exponentially more after it sank beneath the Category 3 storm's churning fury.

This national pattern of throwing good money after bad raises questions about the second supposition undergirding the argument DPW and the *Times* pressed on county supervisors and the public. The newspaper is convinced that the agency alone has the requisite expertise to resolve this issue. Any compromise not guided by their technocratic insights, the *Times* warned, would leave county politicians "playing roulette with nature." Surely that genuflection to its authority cheered DPW, but it left the rest of us a bit rattled.

Years ago John McPhee warned of the peculiar consequences that come from accepting expert knowledge at face value. In *The Control of Nature* (1989), particularly the final chapter, "Los Angeles Against the Mountains," he offers a troubling take on the intense labor (and endless taxpayer funding) required for Angelenos to reside in the creases and folds of the San Gabriel Mountains. No sooner had dams like Devil's Gate been built across canyon mouths than the county began emplacing other flood- and debris-retention structures in the upper reaches of narrow watersheds. To reach them, they built roads that cut sharply into the loose soil, and behind them came housing developers who paved streets and graded and sliced off hilltops, stair-stepping higher into the foothills and canyons. This invasive infrastructure only magnified the volume and rocky rush of debris that, following such conflagrations as the Station Fire, piled up behind Devil's Gate.

Before firing up the bulldozers, clamshell cranes, and ten-wheeler dump trucks to haul off that mass of material, we must take a step back and admit that the dilemma we face is one of our own making. It is framed by our touching faith in scientific experts, our heartfelt conviction that an engineered terrain will solve our environmental conundrums. It has not, and it will not.

That was the conclusion John Tettemer, then acting chief deputy engineer at DPW, reached in a speech in the 1980s. His arresting words could not be more relevant to debate in 2011 over Devil's Gate: "We should stop building things where they do not belong and leave a little room for nature."

Afterword

Homeward Bound

I've always wanted to have an epiphany, but I've never been struck by a blinding flash of life-changing intuition. Perhaps that explains why I doubt its possibility. The fact that epiphanies are staples of environmental confessional literature—in which the author discovers (as does his or her reader) the humbling power of nature or the hubris of humanity—only reinforces my sense that they strike a false note. As contrivances, they are designed to persuade us of a truth their author has already learned but feels compelled to present in a naturalistic formula. Mother Nature has become a trope.

Arguably the most celebrated revelatory moment in American nature writing is Aldo Leopold's tale of how in shooting a wolf he became a new man. In *Sand County Almanac* he recounts spotting a wolf and its pups frolicking along Arizona's Blue River and, full of "trigger itch," began pumping volley after volley into the "mêlée of wagging tales and playful maulings." When Leopold "reached the old wolf in time to watch a fierce green fire dying in her eyes," he wrote, "I realized then, have known ever since, that there was something new to me in those eyes—something known only to her and to the mountain." As for the "hidden meaning in the howl of the wolf, long known among mountains, but seldom perceived among men," he reasserted Thoreau's dictum even as he revised it: "In wildness is the salvation of the world."[35]

The slaughter, Leopold biographer Curt Meine argues, was "burned into Leopold's psyche." There is, however, plenty of evidence that Leopold's compelling story was designed to serve literary ends. In 1944 Albert Hochbaum persuaded Leopold that the manuscript of *Sand County Almanac* lacked an essential affective focus. It "is important that you let fall a hint that in the process of reaching the end result of your thinking you have sometimes followed trails like anyone else that led you up wrong alleys," he wrote. Leopold subsequently drafted "Thinking Like a Mountain," an essay in which he recounts the wolf episode, and which Hochbaum praised, saying it "fills the bill perfectly."[36]

I don't object to Leopold's larger message that we must accept that we are part of a much larger universe in which wolves and mountains are as integral we are. Neither can I dismiss his impulse to make that point through a narrative to which he assigned greater meaning than it carried when he took aim at the gamboling wolves. I know Rev. Canon Kingsley was on to something when he asserted that "history is a pack of lies." Yet I feel manipulated by the studied quality of Leopold's tale, one consequence of which has been to discount the thing he wants me to accept about his fatal encounter and the significance it bore for him and, by extension, us.

More persuasive, it seems to me, is an argument Barry Lopez advances in *The Rediscovery of North America* (1990). Seeking an intellectual counter to the destructive power that defined Spain's New World empire, an imperial arrogance other migrants have adopted as they have run roughshod over lives and landscapes, Lopez argues that we must embrace a long-term schema for understanding who we are *in* nature. "The true wealth that America offered, wealth that could turn exploitation into residency, greed into harmony, was to come from one thing—the cultivation and achievement of local knowledge." To achieve this, one must take up full-time "residence in a place." This is not just a geographical exercise, and it cannot be achieved in a revelatory instant or a climactic flash. Eurekas don't measure up, Lopez suggests. Only by becoming "intimate with the land"—an intimacy that includes knowing its full and complex history—can we "enclose it in the same moral universe we occupy."[37]

Lopez's declarations are more compelling because they heighten our sensitivity to the intricate interaction between nature and civilization, between place and people across time. His insights resonate most in their capacity to guide us to ourselves. Why do we respond to the world as we do? An answer is that past places matter enormously, a concept that figured in the shaping of my biography of conservationist Gifford Pinchot. I realized after years of research that I needed to give full form to his environmental activism, and to do that justice required that I locate him in the physical worlds through which he so energetically moved; his acute awareness of land forms—from his home turf of northeastern Pennsylvania to the American West's terrain to the distant, fecund Galápagos—impelled me to pay attention to what caught his eye.

That work pushed me to think more seriously about my own experiences growing up in Darien, Connecticut, a posh suburb of New York City laid down on top of farmlands that by the mid-1950s were themselves covered with second-growth woods. The rejuvenated forests my friends and I played in were littered with evidence of older economies that had drawn their wealth from the region's stony soil. Tumbledown rock walls marked out the extent and size of former cornfields; orchards shrank beneath the thick canopy of encroaching hardwoods. We turned an ivy-choked, rock-lined cellar into a clubhouse and gathered rusted pails and tin cans to hold other unearthed treasures. Threading though this agrarian mosaic was a network of bridle paths that led to and from the Ox Ridge Hunt Club, a reminder of the grand estates that in the 1920s had absorbed some of the community's smaller colonial farmsteads. By my youth, those holdings had been sold off and carved up for the housing subdivisions and schoolyards in which my fellow Baby Boomers would come of age.

Michael Pollan speaks of encountering these "shadows on the landscape" with an almost haunted reverence. "Oaks, hickories, ash, and sycamores had spread out evenly over the village like a blanket," he wrote after wandering through the ghostly remains of the once thriving agricultural community of Dudleytown in northeastern Connecticut. Trees rose "up in the former yards and fields and even in the middle of cellar pits, jutting heedlessly through spaces that once had been organized into kitchens and bedrooms, warm spaces that had vibrated with human sounds."[38] But what worried Pollan the gardener about this display of nature's dominance, its invasive power, was for my friends and me a matter of secret delight. Here in the ballooning suburbs of the southwestern toe of the Nutmeg State we had stumbled upon a forested redoubt, site of endless games of hide-and-seek (and war). Here we could root around for trinkets that gained value precisely because they had been cast off by people we did not know; here was our midden, a history we could, in every sense, dig up.

I did not become an environmental historian because of my experiences in this archaeologically rich and arboreal playground or my encounters with other captivating landscapes like Claremont, San Antonio, or Los Angeles, seminal though each has been in giving me subjects about which to write, and communities I could write myself into. What fascinated me as a child has continued to enthrall: the local and particular, the palpable, all set within precise moments of time. This sensibility has morphed into an academic agenda freighted with urban streetscapes, plazas, and parks, an agenda containing natural edges (shoreline, riverbank, ridgeline) and human ones (discriminatory zoning, border walls, and dividing highways). To sift through these strata of biography, material culture, and social history; to seek out how others have inscribed themselves on the physical land we now inhabit—these are some of the most enduring ways to recapture the human impulse set within nature's force. They are also keys to establishing a conscientious life in place.

Notes

1. Y-Fu Tuan, *Space and Place: The Perspective of Experience* (Minneapolis: University of Minnesota Press, 1977), 3.

2. Barry Lopez, *Home Ground: Language for an American Landscape* (San Antonio: Trinity University Press, 2006), xviii.

3. Michael Pollan, *Second Nature: A Gardener's Education* (New York: Dell, 1991), 277–78.

4. Penelope Lively, *City of the Mind* (New York: Grove Press, 2003), 68.

5. Lydia R. Otero, *La Calle: Spatial Conflicts and Urban Renewal in a Southwestern City* (Tucson: University of Arizona Press, 2010); Laura R. Barraclough, *Making the San Fernando Valley: Rural Landscapes, Urban Development, and White Privilege* (Athens: University of Georgia Press, 2011).

6. Tuan, *Space and Place*, 6.

7. James L. Nelson, "Business History of the San Antonio Brewing Association (Pearl Brewing Company) 1886–1933" (M.A. thesis, Trinity University, 1976), 239–40. In Milwaukee more than 50,000 citizens gathered on April 7, 1933, to cheer the end of Prohibition in Wisconsin; Bayrd Still, *Milwaukee: The History of a City* (Madison: State Historical Society of Wisconsin, 1948), 493–94.

8. *San Antonio Light*, Sept. 15, 1933, 1A; *San Antonio Express*, Sept. 15, 1933, 1A, 4A.

9. *San Antonio Light*, Sept. 15, 1933, 1A. Beer advertise-

ments—Pearl's was a full-page spread—blanketed pages 12A–23A. Making beer legal did not revive one brewery tradition: "The old brewery horse has passed into history," the *Express* mourned; "the spirited heavy draft animals that once drew the huge brewery wagons through the street before Prohibition have given way to trucks" (Sept. 15, 1933, 4A). Milwaukee and Cincinnati recorded leaps in brewery production and employment; Still, *Milwaukee*, 393–94; William L. Downard, *The Cincinnati Brewing Industry: A Social and Economic History* (Athens: Ohio University Press, 1973), 134–38.

10. B. B. McGimsey to distributors, March 30, 1933, quoted in Nelson, "Business History," 232.

11. Anthony Gevers to B. B. McGimsey, April 24, 1933; B. B. McGimsey to Anthony Gevers, May 9, 1933, quoted in Nelson, "Business History," 233–34; *San Antonio Express*, Sept. 15, 1933, 1A, 4A; *San Antonio Light*, Sept. 15, 1933, 1A, 8A; Sept. 14, 1933, 7B; 1B. This essay focuses on Pearl Brewery, but I am mindful that Lone Star Brewery, started in 1884, was an important player in the city's economic life. *San Antonio Express*, Sept. 18, 1884, in Donald E. Everett, *San Antonio: The Flavor of Its Past, 1845–1898* (San Antonio: Trinity University Press, 1976), 138. Just how important brewing was to Milwaukee is evident in employment figures. Within a year of repeal more than 8,000 were employed in the city's nine breweries and allied payrolls; Still, *Milwaukee*, 393–94.

12. Char Miller, "Where the Buffalo Roamed: Ranching Agriculture in the Urban Marketplace," in *On the Border: An Environmental History of San Antonio*, ed. Char Miller (University of Pittsburgh Press, 2001), 56–82; Fred Mosebach quoted in Travis E. Poling, "Area's heritage soaked in beer," *San Antonio Express-News*, Nov. 24, 2002, 1K; Char Miller and David R. Johnson, "The Rise of Urban Texas," in *Urban Texas: Politics and Development*, ed. Char Miller and Heywood

Sanders (College Station: Texas A&M University Press, 1990), 14–17, 28–29.

13. Nelson, "Business History," 8–11.

14. James D. Ivy, *No Saloon in the Valley: The Southern Strategy of Texas Prohibitionists in the 1880s* (Waco: Baylor University Press, 2003), 7–23, 103–4. San Antonio strongly opposed the 1887 statewide referendum to enact Prohibition; the final tally was 4,861 opposed, 507 in favor.

15. Frank Bushick, *Glamorous Days* (San Antonio: Naylor Company, 1935), 32; Judith Berg Sobré, *San Antonio on Parade: Six Historic Festivals* (College Station: Texas A&M University Press, 2003), 114. Downard (*Cincinnati Brewing*, 164–68) illustrates the German impact on the Queen City.

16. *San Antonio Herald*, Feb. 23, 1877; *San Antonio Light*, Aug. 25, 1886, quoted in Nelson, "Business History," 4. The saloons in Little Rhein differed from their competitors throughout the city. "Theirs are not fiery drinks, but cooling, filling lotions; beer, weiss beer, and more beer." But the setting was decidedly domestic, the *San Antonio Express* affirmed on Aug. 17, 1897: "When you enter [them], you must tread lightly, and cautiously pick your way over and around toddling infants, pans of half-peeled potatoes, or a long stretch of calico that the housewife is fashioning into a house wrapper, for a South Alamo street saloon is nothing if not intensely home-like. You walk up to the counter and you can look squarely into the living room"; quoted in Everett, *San Antonio*, 45–46.

17. Bobbie Whitten Morgan, "George W. Brackenridge and His Control of San Antonio's Water Supply, 1869–1905" (M.A. thesis, Trinity University, 1961), 98.

18. I am grateful to my talented research assistant, Elizabeth Adams, for her critical research in the Sanborn maps and San Antonio City Directories.

19. Poling, "Area's heritage," 1K; Elaine Wolff, "Strange brew," *San Antonio Current*, Oct. 30–Nov. 5, 2003, 10.

20. The German presence in breweries throughout the United States was marked; see Downard, *Cincinnati Brewing*, 10–11, 64–68, 222; Still, *Milwaukee*, 188–89.

21. Nelson, "Business History," 136–37.

22. Sobré, *San Antonio on Parade*, 51–72; see also Everett, *San Antonio*, 52. An all-black Fireman's Union was organized in the first years of the twentieth century and proved immediately successful in shortening its members' work hours without a decrease in pay; Nelson, "Business History," 145–46.

23. Nelson, "Business History," 123–60. San Antonio's brewery workers' salaries and hours are consistent with those earned in Cincinnati and elsewhere; Downard, *Cincinnati Brewing*, 97–121.

24. Wolff, "Strange brew," 10, 12; Nelson, "Business History," 129–36, 141, 144, 154–55; Poling, "Area's heritage," 1K.

25. Stanley Baron, in *Brewed in America: A History of Beer and Ale in the United States* (Boston: Little, Brown, 1962, 340–46) notes that transformation of the brewery business was consistent with alterations in railroads, hotels, and airlines; beer advertising nationwide accounted for $6 million in 1938 and an estimated $95 million in 1960.

26. Wolff, "Strange brew," 12; Scott Huddleston, "S.A. suds' stories," *San Antonio Express-News*, July 8, 2001, 2B; Travis Poling, "Brewery over a barrel again," *San Antonio Express-News*, July 27, 2002, 1E; Travis Poling, "Brewery's fate on tap," *San Antonio Express-News*, May 17, 2000, 2E; Sanford Nowlin, "Plant a part of brewing history," April 22, 2000, *San Antonio Express-News*, 13A.

27. Quoted in Wolff, "Strange brew," 10, 12.

28. www.americanrhetoric.com/speeches/jfkhouston ministers.html, accessed Feb. 9, 2007.

29. www.jfklink.com/speeches/jfk/sept60/jfk120960 _alamo.html, accessed Feb. 9, 2007.

30. www.jfklink.com/speeches/jfk/sept60/jfk120960_houston02.html, accessed Feb. 9, 2007.

31. www.americanrhetoric.com/speeches/jfkhoustonministers.html, accessed Feb. 9, 2007.

32. Holly Beachley Brear, *Inherit the Alamo: Myth and Ritual at an American Shrine* (Austin: University of Texas Press, 1995), 1.

33. Randy Roberts and James S. Olson, *A Line in the Sand: The Alamo in Blood and Memory* (New York: Free Press, 2001), vii.

34. Richard R. Flores, *Remembering the Alamo: Memory, Modernity, and the Master Symbol* (Austin: University of Texas Press, 2002), 60.

35. Aldo Leopold, *A Sand County Almanac* (New York: Oxford University Press, 1949), 129–33.

36. Curt Meine, *Aldo Leopold: His Life and Work* (Madison: University of Wisconsin Press, 1988), 93–94, 455–59.

37. Barry Lopez, *The Rediscovery of North America* (Lexington: University Press of Kentucky, 1990), 21, 29, 31, 37.

38. Pollan, *Second Nature*, 36–37.

Acknowledgments

As ever, my mother almost gets the last word. She taught me to read and so much more, and to her memory this book is lovingly dedicated. She would understand why she shares that honor with Samuel Cassidy Miller, our first grandchild, whose birth she anticipated but did not live to celebrate. Putting them together in print is as close as they will ever get, alas.

On the Edge owes its origins to many kind editors who have allowed me to publish in their journals, magazines, newspapers, and online sites. They let me test out my ideas and arguments in the form of essays, commentary, and memoir. I have rewritten each one for this book, but I know my ability to do so is predicated on their generous willingness to let me write for them and to their readers in the first place. I especially want to thank Zach Behrens, Mike Brossart, Lou Dubose, Greg Harman, Alejandro Manrique, Hernando Ramirez, Ellen Shull, Steve Taylor, Julia Wells, and Elaine Wolff. David Taylor at the University of North Texas was instrumental in helping me see how all this prose could become a book. Longtime friend and colleague Barbara Ras—and her talented crew at Trinity University Press, Sarah Nawrocki, Tom Payton, and Burgin Streetman—has been as essential to the development of this collection as she was on three previous books. Then, as now, she has been a dream to work with.

As gracious have been my family and a host of friends and colleagues who listened, perhaps more carefully than they should have, to my babbling about the stories and incidents and moments that make up many of these narratives. In San Antonio I learned more than I can say from Vicki Aarons, Anene Ejikeme, Ted Flato, John Garland, David Johnson, Gary Kates, Jed Maebius, John Martin, Alida Metcalf, Nancy Mills, Barbara Ras, Linda Salvucci, Woody Sanders, Bob Sohn, and Mary Kay Stewart, among so many others. In Claremont my remarkable colleagues and amazing students in the Environmental Analysis Program have shown me how to make the transition to new ways of thinking, teaching, and writing.

And above all, to Judi. We met in Claremont and since then have moved to Baltimore, Ithaca, Miami, and San Antonio before returning to the place where we began our life together. Everywhere and in every way, she has turned house into home.

Char Miller is director of the environmental analysis program and W. M. Keck Professor of Environmental Analysis at Pomona College. He was formerly a professor and chair of the history and urban studies programs at Trinity University. He is the author of *Gifford Pinchot and the Making of Modern Environmentalism, Deep in the Heart of San Antonio: Land and Life in South Texas,* and *Public Lands/Public Debates: A Century of Controversy,* and the editor of *On the Border: An Environmental History of San Antonio* and *Fifty Years of the Texas Observer.* Miller writes a weekly column for KCET Los Angeles focusing on environmental issues confronting the West. He lives in Claremont, California.